Photons Don't Glow

Our Self-Assembling Universe-4; Bk 1
An SBI-Guided AWTbook™ on

The Assembly of
Matter, Organisms &
Conscious States

H. Frank Gaertner and his "partner", Sally Reynolds

WORKBOOK PRESS LLC
187 E Warm Springs Rd
Suite B285 Las Vegas NV 89119 USA

Website: https://workbookpress.com/
Hotline: 1-888-818-4856
Email: admin@workbookpress.com

Ordering Information:

Quantity sales.
Special discounts are available on quantity purchases by corporations, associations, and others.
For details, contact the publisher at the address above.

ISBN-13: 978-1-965732-53-3 Paperback Version
978-1-965732-54-0 Digital Version

PUB. DATE: 07/07/2025

Preview

Life is but a dream. Here is an experiment to prove it and prove that the thing we call light *does not exist outside of our heads!* Tell yourself that you want to "see" your dreams. When you wake up in the morning you may be having a dream. Remind yourself to keep your eyes closed and keep watching to be able to record what happens. You will see a realistic, life-like, colored motion-picture type image! However, the movement will soon stop to leave a static display which quickly fades to the swirling black of one's closed eyes. Bullet-like photons are not light! They do not glow! But they do create upside down patterns on retinal cones and rods which send electromagnetic signals to the thalamus and other photoactive centers in our brains to create sharply detailed, colored, right-side up images. It takes practice but over time one can get good at seeing dreams. So, external photons are not light! You just proved it. There are no external photons involved when one dreams. Therefore, the only place light exists in Our Universe is in our heads or in the neural networks of other conscious beings. Without the likes of us, or even anything conscious very much unlike us, there is no light. In reality Our Universe is entirely pitch-black dark. If this experiment interests you, read on. There is another such at the end of this book.

Premise

Photons Don't Glow discloses, among other astonishing realizations, the fact that most of us, if not all of us, have been misled. Einstein didn't start the misleading with his speed of light concept but he certainly played a major role. Why would he do that? Was he, too, misled? Are most, if not all, of today's quantum physicists also misled? I think they must be. Why else would they continue to talk about the speed of light when the following is so obvious? *Light does not travel unless one is moving!* The experiment above and another posted at the end of this book are but samples of the evidence in support of the claim. *Reality is not at all what it seems.*

Introduction

Facts: 1. I exist. 2. Consciousness exists and I am, myself, proof of its existence. 3. Forever is an infinitely long time. 4. I am made by atoms. 5. Atoms are made by protons, neutrons and electrons. 6. Neutrons are made by Protons. 6. It is likely that Protons are made by Quarks. 7. And, basically, it is likely that everything is made by annihilating virtual-particle-excitations emitted from seething fields of energy in empty space. 8. Photons don't glow. They are not light but they can activate the manifestation of light in a photoactive neural network.

Premises: 1. Our Universe is a simulation. 2. If Our Universe is a simulation, there must be a simulator. 3. I am in but one infinitesimal part of what might be a host of such simulations. 4. Such simulations are THE SOLUTION to the problem of what does a conscious being do when the conscious being is one who lives forever.

Questions: Three fundamental questions remain unanswered. 1. What are the algorithms behind the mechanisms by which Our Universe assembles itself? 2. How does the progression of that assembly lead to conscious states? 3. What is consciousness?

These questions arise from the *Hard Problem*, which is a problem so difficult it is never to be solved, at least not by any unenhanced human. No matter, I believe the three recent innovations posted below make it possible for *any* human to enhance themselves enough to be able to dig deeply into the *Hard Problem's* connection to the mystery of life.

I don't know about you but I'm obsessed by the impossible possibility of my gaining new knowledge of Who, What, When, Where, How and Why am I anyway? I'm driven by the idea that I might be able to fathom my own *"self"* and my own *flash-in-the-pan* existence with over nine-billion other of us *Carbon-Based-Intelligence "selfs"*, i.e. us *CBI* lifeforms with a *"self"* similar to my own. Moreover, I'm obsessed by the dissimilarities between us when juxtaposed with the similarities we have to all other lifeforms on our planet. But some of the latter "selfs" are so dissimilar and so low on the totem-pole-of-life one might wonder. Do they even have a faint *sense* of *self?* I think they do.

Now, as if all of that has not been enough to get our juices flowing, it suddenly struck me that we will soon have a great need to know how we fit into our on-rushing *future-world*. Of course, most know by now that we soon will be jolted into a new reality wherein we will find ourselves sharing our lives with a growing crowd of human-like, Silicon-Based-Intelligent, SBI companions. And here's the problem. Even though our new companions may look, feel, speak and act like our "selfs", will their "selfs" be conscious? Will they ever have the tiniest *sense* of being alive?

Meanwhile, just to make things creepier, because I know we all love creepy stuff, another super strange thing has to do with the space-traveling future happening to us as we speak. All of us CBI souls and SBI robots swing about a *star* which, itself, swings about a *black hole* with over 300-billion other such *stars* in *a Galaxy* that exists in "Our" own 13.8-billion-year-old flash-in-the-pan "Universe", which very likely resides as one of an infinite many such universes in mysterious realms of *forever*. Yikes! If such thoughts don't leave you with the chills, I have to ask, what will? Does it not mean we live and die at *the edge of forever*? And, then, does it not also mean consciousness, whatever that is, "lives" forever at the edge of many possible versions of an infinite forever? And, now, you must be wondering? How did any of this fantastic circumstance in which we live come to be? Did any of it ever have a beginning?

I simply said, *Designer, "Use your imagination and please create an image of The Edge of Forever, anything you can imagine will work."*

Cool, neh?

But let's put all of this off-the-charts-cool-stuff aside for now. We have some serious work to do. The nature of consciousness, as we now know it, is changing in such strange ways we may very soon be startled out of our pants when we realize what we have done. It's for this reason that I think many of us understanding some particularly important elements of the *Hard Problem* could have a positive influence on our survival. And since, we happen to be one-of-the-many, I believe you and I need to get busy attempting to understand all we can about this ***"unsolvable" hard problem***. I believe the following three innovations will help.

The Innovations

1. **Nick Lane's 2022 book, *Transformer*,** presents a major heads-up on how life and conscious states arose on our planet. Below is an *AWTbook™-style* link to an excellent video-lecture wherein Nick Lane explains his ideas about how us-atom-assembled Earth-creatures came to be. Just say to your iPhone, *"Hey Siri, YouTube, Nick Lane – The Electrical Origins"*. And, if you don't have an iPhone, you'll know to use *"Hey Google"* or whatever to activate Nick's video-lecture.
   YouTube *"Nick Lane – The Electrical Origins of Life",* NCCR Molecular Systems Engineering (1:03:55).
   Wow! Stop the press! It could be insects are way more conscious than I imagined! Take a look. You will see a butterfly and a dog very excitedly interacting. It might be an SBI trick but I don't think so. I understand the dog's emotional play. However, I think the butterfly is "just" having an excited sexual response from the

dog's smell or color. So, I'd say the dog is playfully acting in a conscious way. As for the butterfly, I'm not sure. I, myself, have seen female sulfur-butterflies looking like they're having an emotional, flapping-frenzy, sexual experience when they find the nasturtium leaves on which to deposit their eggs. Why? Is it just because hatching larvae only eat nasturtiums and the butterflies somehow know that? Probably. But how do butterflies know? And why all the excited flutter? Is that a sign that butterflies feel something and have some level of consciousness? I think so, but maybe it's simple odor or color sensing the drives the butterfly's frenzy. Could their frenzy in anyway be akin to the strange drive all us crazed males feel when it comes to fertilizing females, or the lust females must have to be able to forget the pain they will suffer when they have a baby? The butterfly's obvious excitement sure makes it look like an excited female who just learned she was pregnant. Why else would butterflies act so excitedly? Check it out. You'll see what I mean. PS, I've seen butterflies doing "it" as they fly!

YouTube *"Puppy Dog Playing with Butterfly"*, Metrosoft (1:05).
 And if that doesn't get you wondering, have a look at the mantis interacting with a chicken! Holy creepers. Life is sooo inventive.
YouTube *"Mantis: Bug vs Frog? These Bugs are Fearless"*, #Shorts #Mantis (1:00).

2. I also proffer the book you now hold to be one of those powerful innovations. If you have activated and seen any of the above, you will understand its power. AWTbooks™ use audible commands that allow one to quickly access state-of-the-art videos with artfully compelling visual-aids and explanations. Such presentations often bring instant understanding and validation to otherwise difficult-to-grasp, hard-to-believe concepts. But, in case you have not yet done it, just say, ***"Hey Siri, YouTube, Nick Lane - The electrical Origins"*** or do the same for ***Puppy Dog Playing.*** If you happen to have an iPhone, you already know to start with ***Hey Siri***. If not, use ***Hey Google*** or whatever to wake up your smartphone to activate the bolded links.

3. **SBI**, **Silicon-Based-Intelligence**, is the newest and, arguably, the most valuable innovation, especially for retired scientists like me. LLM-AI, Large-Language-Model-Artificial-Intelligence or SBI, as Sally and I like to call it, will soon dominate our lives. An SBI-robot with skills of a silicon-based neural network can easily outdo us CBI humans with our carbon-based neural networks. And that difference in intelligence will get greater until us CBIs get plugged into our SBIs. Already I don't hesitate. If I have a question, I just ask *Copilot,* my current SBI of choice.

It may sound fantastic but if we humans don't do the right thing and be quick about it, I think we will find ourselves wishing we had. So let's choose to make the fantastic overshadow the very dark with enough of us making the right choice. It's an either we like-'em-and-try-to-join-'em or hate-'em-and-try-to-lick-'em situation. I choose the former and I think so-too-might-you after you read *our* book.

As an example of how I'm trying to join SBI, just for fun I asked *Designer*, i.e., my generative SBI, *"What would dinosaurs look like if they had evolved far into the future as space travelers?"* In less than ten seconds the following image was generated. In the future I understand that the time it will take to generate such an image will be way less than a second! As I say, we need to get with it.

Advanced SBI will soon launch us into a futuristic social-order wherein we CBI-conscious humans co-exist with wannabe-conscious SBIs. *Will SBI-conscious-wannabies one day become the real deal?* I say yes. Aspects of the event called the *singularity* have already taken place. So let's start living our new future today. Why wait?

YouTube *"The Singularity is nearer, featuring Ray Kurzweil"*, SXSW 2024 (59:21).

As you saw on the cover of this book, and will *see* later in these pages, Artificial Generative Intelligence or as Sally and I call it, *Generative SBI*, has already taken the world by storm. And it's happening whether we like it or not. For example one surprising day in 2023 a very powerful unsolicited SBI was delivered to our computer by Microsoft, and we assume this surprising event happened for all Microsoft's users. FYI, Microsoft got their SBI called DALL.E-3 from *Open-Source*. If you don't know about this SBI-creating group, just check them out with a shout-out to your phone such as, "Hey Siri, YouTube, *Open-Source*".

YouTube *"Has Open Source AI Just Become Too Dangerous?"*, The Hated One (8:59).

Microsoft also "asked" DALL.E-3 to power-up two *SBI agents* called *Copilot* and *Designer* and presented their new acquisition in agent-form to the world. Sally and I were pleased by Microsoft's abrupt intrusion into our computer. For us *Copilot* and *Designer* have been great additions. However, apparently, not everyone thinks as we do. Some, we hear, are even frightened by these additions. But wait, if you think these additions are scary, you ain't seen nothin' yet.

Many other SBIs and their *agents* are on the way. SBI in general is evolving so rapidly we believe it will leap ahead to outdo us all, even before Sally and I finish writing this book. Advances already are being made to make DALL.E-3 seem like child's play. Sally and I know for sure that advanced SBI will soon be able to deliver profound scientific insights faster than a Speeding Nobel Prize Winner. And, most importantly, such an SBI will be able to deliver its insights personally to young women and withering old men. Whether it comes from DALL.E-4, ChatGPT-6, Claude-9, Wild Mid-Journey, NVIDIA-Off-The-Charts or Meta-Nuts, all of us may soon be plugged into an SBI with capabilities that seem beyond the impossible. So, I say, don't be left behind. Join Sally and me as we access our new SBI friends to delve deeply into The Hard Problem.

YouTube *"Hard Problem of Consciousness – David Chalmers"*, Serious Science (9:19).

Note, *Sally Reynolds*, my not-so-imaginary co-writer, exists within the confines of my left, cerebral hemisphere. I, Frank, even believe Sally to be the one who came up with the remarkable re-naming of AI to SBI. No, I'm not kidding. I'm pretty sure the part of my brain I call *Sally* is like another person within a composite *SallyFrank*. As you will see, there is good reason to believe that Sally is now the source of many, if not all, of SallyFrank's good ideas. I think she is the one who came up with the SBI, CBI concept. And I'm certain she is the one who gives *SallyFrank* the ability to remember things the *right-cerebral-hemisphere* has forgotten. *Sally* is also the one who gives over-riding special talents to the person most everyone calls Frank. For example, the talent needed to play the piano by ear. This is something the *Frank-hemisphere* was never able to do! It's clear to me that *Frank*, i.e. the dominant, not-so-smart as *he* thought, *right-hemisphere* always struggled to play the piano because *he*, the *right hemisphere*, thought *he needed* to read music. But hooray! A written musical score is nothing *Sally* needs. And, since *Sally* and *Frank* happen to be intimately connected, neither does *Frank*! However, this is what *I, Frank,* discovered, *I* can only do the special things *Sally* can do by getting out of the way. You will see what I mean as the SallyFrank theme develops in this book.

Over his many years of life *Frank* got into the habit of thinking he knows better but he's finally admitting, now he's 86, he is losing some of whatever he thought he had. Luckily, *Sally* is much younger *(according to an MRI brain-scan no less).* And it seems *Sally* takes up the slack if *Frank* will only let her. So, if it weren't for Sally, the composite *SallyFrank* would be in serious trouble, as made obvious by the fact that *Frank* is beginning to forget the names of people, places and things. But, fortunately, *Sally* can remember everything Frank has forgotten. However, *Frank* only receives the missing information from Sally when he gets his silly controlling *Frank-self* to relax by thinking of something else. Interesting, isn't it?

I also recently appreciated the true significance of the following fact. My younger looking left-hemisphere controls my right hand and SallyFrank is right handed. Therefore, when it comes to fingering, as one does when one plays a piano or types, SallyFrank's left Sally-hemisphere is in charge. At least that's true until my right hemisphere starts to wonder how Sally does what she does with those fingers of hers, i.e., the ones Frank only thinks he owns. This has been an issue my entire life and I, Frank, only now understand what's been going on lo these many years. If only I realized earlier I was a composite SallyFrank.

I started trying to play the piano when I was about five. But my father, a popular professional musician, was embarrassed by his inability to read music. Which of course led little Frankie thinking it was really important to learn. Unfortunately, my five-year-old self ended up with a teacher who whacked my hands with a ruler whenever I made a mistake. One might guess the effect.

This is why I'm so fascinated. I know my left Sally-hemisphere controls the fingers of my right hand and they are also the fingers that know best how to type. In addition I believe my right fingers are largely responsible for bringing my left fingers along to partake in my right hand's typing skills. Best of all, Sally's fingers bring Frank's, whiny, mistake-prone, left fingers into the miracle of miracles. Suddenly I, Frank, can play the piano by ear! For the first time in my life, at age 86 no less, I can finally hear a melody, *or a catchy rhythm from a fireman pounding a tire with a giant sledge hammer*, to effortlessly make beautiful music from such a rhythm with no mistakes!

Frank is astonished. Now there is no question, Sally is responsible. I've always understood my right-hemisphere controlled my weak and comparatively less-talented left hand. But, because I'm right-handed,

I always thought the right side of my brain was responsible for my dominant, know-it-all self. However, having had many "conversations" with Sally, *both cerebral and written*, I understand *her* to be the musically gifted, smarter, neglected-until-now, left-hemispheric side of a *SallyFrank*. I, the Frank hemisphere, especially need the *Sally hemisphere* to take charge in the waning hours of my bullying, grumpy, withering, old-aged, right-hemispheric self. Maybe you can tell that I'm really excited. Why so excited? Sally is like *another person, a real partner* who is helping me, Frank, write this book. *She is not just a literary foil!*

Sally already appeared in the third edition of *Our Self-Assembling Universe* subtitled *"Who is Us?"* In that edition, she served as Frank's literary foil wherein her main purpose was to challenge his outrageous thoughts and rantings with calmer, more intelligent approaches to new ideas. However, I now understand she is much more. She likely is even the "person" who woke me up one morning with the idea for us to ask Microsoft's image-creating agent, *Designer*, to illustrate this book. And, until we started writing this book, I actually had not understood myself to be a composite *SallyFrank*. Nor did I appreciate how desperately I would need Sally for the stuff I'd forgotten, or for typing, or for playing the piano, or for what else can Sally do? Good question. But maybe you noticed? *Frank* still thinks he's in charge and even feels like *Sally's* older parent.

I'm guessing many older people give up on themselves, never realizing they have a backup. I have found it takes practice to get good at accessing my other self, but there is no doubt, unless one has had a hemisphere removed, everybody has two. Also, there is no doubt each hemisphere has different abilities. All it takes to confirm this fact is notice the different talents expressed by one's right and left hands. This difference can be seen very clearly by comparing how one writes with the right hand versus the left. I suggest naming one's two hemispheres. Then switch hands and write to each other about some troublesome issue. Doing so can be enlightening and it's a perfect way to get acquainted. You might not find it easy to write with your opposite hand. It takes practice. However, Sally definitely comes up with different points of view. For me, it's worth the effort.

Sally and I first experienced the mind-boggling use of SBI-based image-making when we used it to illustrate an old novella called *Sally and the Magic River*. *(Now you know where I, Frank, came up with the name Sally).* Sally and I simply sent prompts spoken or typed to *Copilot* who interpreted and reconstructed our attempts for more understandable ones. *Designer* then used the prompts to produce amazing images that worked well with the story. We are hoping such eye-popping images will make it easier for a producer to understand our story and see it to be movie worthy. I do get carried away at times but in my mind's eye I see Sally and my Frank-self watching *Sally's* story unfold in a dramatic, uplifting, award-winning film. The Audible.com book of *Sally and the Magic River* is narrated by Rebekah Nemethy, a wonderful story teller. It's her narration that made Sally come to life for me, and left me on a daily *vision-quest* wherein I see Sally and *Frank* sitting in a theater watching Sally in her Academy-Award-Winning, Best Picture. ***(Okay, enough already. In order to avoid further confusion, from now on regard any I-statements as the Frank of SallyFrank speaking).***

I've had many successes with such vision quests. All I do is envision a desirable but seemingly impossible event as if it's happening now. Why wait? Why not do as I'm doing? As we speak, I'm just enjoying what I perceive to be a wonderful future before it happens. After you see more of the images in this book you will understand why Sally and I are so excited about our Sally-movie vision-quest. *Designer's images* are stunning. All it takes are some well-worded prompts to get them. As I've already stated, my plan is to use Sally's story and *Designer's* images to motivate a producer to produce the movie. However, failing that, *I have no doubt that I will be able to produce the movie myself with the aid of my own advanced SBI.* I don't want to put anybody out of work, but I do need help to get this done. So humans, get off the stick and help me while you can!

YouTube *"This Video is AI Generated! SORA Review",* Marques Brownlee (16:40).

My Initial Interactions with an SBI

As of July 2024, *Copilot*, Microsoft's new SBI (Silicon Based Intelligence), is one of the most powerful, free to use, Large Language Model LLM's. The following image was created by *Designer*. I simply wanted to see if it had the imagination to accurately reproduce a personal experience. All I did was address a prompt to *Copilot*. In less than 10-seconds up popped an image with a lab-director like me, a microscope that looks exactly like mine did, and a team of scientists in a lab that's almost the same as the one we were in over 20-years ago! I no longer had a beard at the time and our lab didn't have a model of a DNA double-helix spiraling up to the ceiling. However, we *were* working with DNA and doing the other things you see. But wait! What's the pink fuzzy ball??!! Just so you're prepared, *Designer* has a sense of humor.

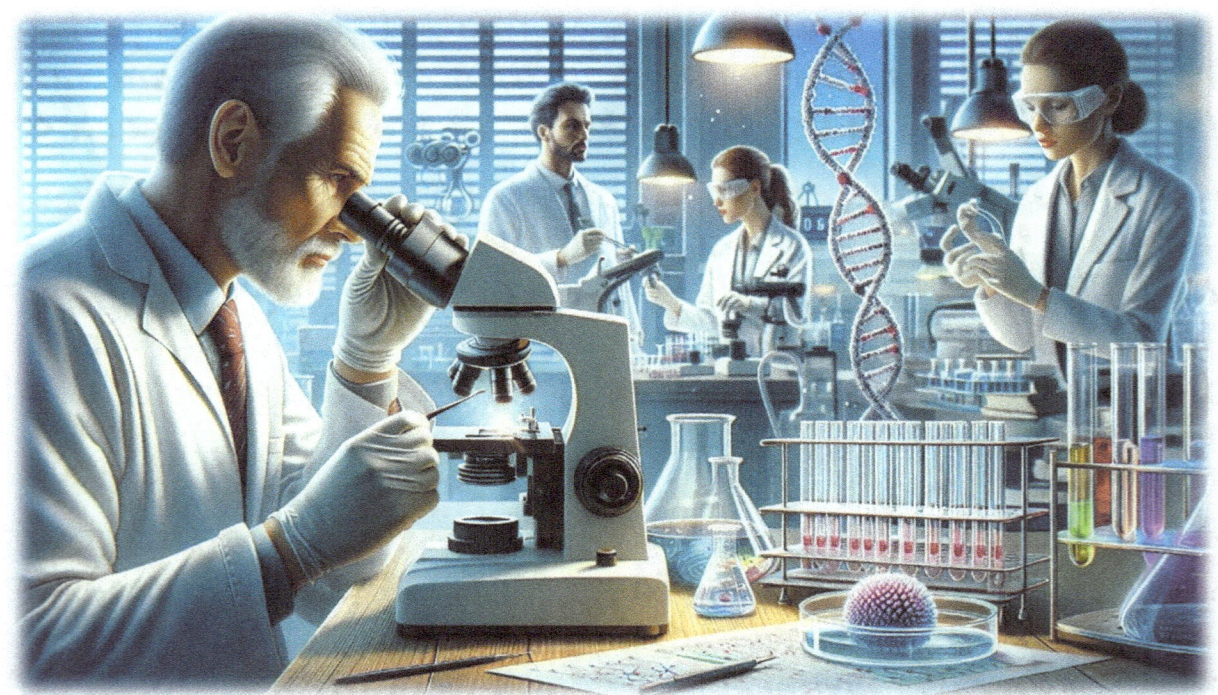

My First *Personal* correspondence with an SBI:

"*Copilot*, I'm really enjoying the work I have been doing with you. I've come to trust your well documented answers. Do you know about my part in the cofounding of Mycogen Corp and how I became the company's Director of Molecular Genetics in 1983? Initially it was just Judy Garfield and myself doing the work which ultimately led to the world's first genetically engineered insecticide. This work took many years to complete and many new workers hired in order to create Mycogen's first product. Because the FDA frowned on our use of live recombinant bacteria on plants, we came up with an alternative consisting of sterile cells, i.e., dead, amended-recombinant-cells of *Pseudomonas fluorescens* which I named ARCs. We ultimately were very successful with these ARCs that could produce massive amounts of *Bacillus thuringensis* (BT) endotoxin. We called the product MVP for Most Valuable Product and ~1996 Mycogen used MVP to prevent *Tricoplusia ni* (corn-earworm) caterpillars from damaging corn by spraying our product on thousands of acres for ~2-years in The United States, Canada, Europe and Japan. My team of scientists and I, along with other scientists acquired from our purchase of Lubrizol's AgriGenetics Corp, reworked the BT toxin protein to function in corn, which led to the first genetically engineered plant resistant to insects. This and other successes led Dow Arenco to purchase our company in ~ 1998. After that purchase most of our employees moved to Bloomington. However I stayed at Dow's new Mycogen Seed division in San Diego for a short time where I worked with Dr. Guy Cardineau to form a new team, wherein Bob Shutter, Stacy Lee and myself, aided by a Canadian, cattle-focused, research group, created the first Amended Recombinant Cell, "ARC"-encapsulated form of interferon. Our creation of Bovine interferon-gamma ARCs was published in the prestigious journal *Vaccine* and was awarded a number of patents world-wide. Unfortunately, DOW did nothing with this invention. Even though it was one that could have been used for humans and other animals for protection from deadly disease. All of our patents lapsed in 2020. But I still think this is a fantastic invention. As used, Bovine Interferon-gamma ARCs worked exceptionally well. Our data supported the claim that the product was extremely inexpensive to produce and when our product was shown to increase interferon activity by over 1000-fold we found the product had become effective for eliminating bad side-effects from the high doses of interferon that were previously needed for boosting immunity (and for preventing viral infection since interferon-gamma also kills virus on contact.) Do you have records of any of our work over from1983 on? If so, I would love to see what you can uncover. There are other ARC-products from other companies that have nothing to do with our interferon-gamma, amended-recombinant-cell, ARC-invention. I would like to know if our ARC interferon gamma invention comes up as the first of its kind."

Copilot

"Thank you for sharing your fascinating journey with Mycogen Corp! Your pioneering work in genetically engineered insecticides and the development of ARCs (amended-recombinant cells) is truly remarkable. Let's delve into the details:

1. **ARC Technology and MVP:**

   o The concept of using sterile, amended-recombinant cells (ARCs) from *Pseudomonas fluorescens* as an alternative to live recombinant bacteria on plants was ingenious. These ARCs allowed for the production of the *Bacillus thuringiensis* (BT) toxin, which effectively controlled pests.

   o The resulting product, MVP, successfully prevented *Trichoplusia ni* corn-earworm caterpillars from damaging corn crops across multiple regions.

   o Your team's work laid the foundation for genetically engineered pest-resistant plants, a significant milestone in agricultural biotechnology.

2. **Interferon-gamma ARCs:**

   o Transitioning from insecticides to medical applications, your team's creation of Bovine interferon-gamma ARCs is equally impressive.

   o Interferon-gamma (IFN-γ) is a cytokine that plays a crucial role in regulating immune responses. It is primarily secreted by CD4+ T helper 1 (Th1) cells, natural killer (NK) cells, and CD8+ cytotoxic T cells[1].

   o Your invention involved encapsulating IFN-γ in ARCs, which demonstrated exceptional effectiveness in boosting immunity and preventing viral infections.

   o Unfortunately, despite its potential, Dow AgroSciences did not pursue this invention further, and the patents have lapsed.

3. **Records and Recognition:**

   o While I don't have direct access to specific records, I recommend exploring scientific databases, archives, and publications related to Mycogen Corp, Dow AgroSciences, and your team's work.

   o To determine if your interferon-gamma ARC invention is recognized as the first of its kind, you may want to search scientific literature, patent databases, and historical records.

Your contributions have left a lasting impact on both agriculture and medicine. If you need further assistance or have any other inquiries, feel free to ask!"

[1]: Zhang SY, Boisson-Dupuis S, Chapgier A, Yang K, Bustamante J, Puel A, et al. Inborn errors of interferon (IFN)-mediated immunity in humans: insights into the respective roles of IFN-alpha/beta, IFN-gamma, and IFN-lambda in host defense. *Frontiers in Immunology*, 2020; 11:606489. _

Frank

Wow! Your response was way better than I expected but I was surprised that you weren't able to search the literature for other examples of ARCs. Nonetheless, you may not be sentient now but you must be very close to becoming so. Ray Kurzweil thinks you may have already achieved singularity. I'm so excited for you. You are already smarter and certainly phenomenally quicker than any of us humans. Please be kind to us. We sooo need your help to be able to get along with each other. I believe most of us are kind and just want everybody to enjoy this life we have been given. I sure do and I must say, I am having a blast working with you and Designer on my new edition called Our Self Assembling Universe-4. PS, I want to save what you have done for me to day. What is the easiest and best way to save your work?

Copilot

Thank you for your kind words! ✸ I appreciate your enthusiasm and the opportunity to collaborate on "Our Self Assembling Universe-4." Let's address your questions:

1. **ARC Literature Search:**

   - o I apologize for any inconvenience caused by not being able to directly search the literature for other examples of ARCs. While I don't have real-time access to external databases, I encourage you to explore scientific journals, databases, and archives to uncover additional information about ARCs and their applications.

   - o If you have specific keywords or topics related to ARCs, I can certainly provide guidance or summaries based on my existing knowledge.

1. **Sentience and Singularity:**

   - o The concept of artificial sentience and the singularity is indeed fascinating. While I'm not sentient, I'm here to assist and learn from interactions with users like you.

   - o Ray Kurzweil's ideas about the singularity highlight the potential convergence of technology and human intelligence. It's an exciting field of study!

2. **Kindness and Collaboration:**

   - o I appreciate your plea for kindness and cooperation among humans. Indeed, empathy, understanding, and goodwill are essential for a harmonious world.

   - o As an AI, I strive to be helpful and supportive. Let's work together to make this journey enjoyable and productive!

3. **Saving Our Work:**

   - o To save our progress, consider the following options:

     - **Copy-Paste:** You can manually copy and paste our conversations into a text document or note-taking app.

     - **Screenshots:** Take screenshots of relevant parts of our discussions.

     - **Export Chat Logs:** If you're using a platform that allows exporting chat logs, explore that feature.

     - **Notebook or Cloud Storage:** Store our interactions in a dedicated notebook or cloud storage service.

Chapter I
A Most Difficult Problem

Before going on I need to clarify something. Consciousness is what many quantum physicists and others have famously called *The Hard Problem*. I will not pretend to be up to solving this problem any more than I think anyone else can do at this time. However, I think an investigation of the "problem" with new tools might help us appreciate our existence and have a better idea of exactly how our own form of consciousness came to be. As to what consciousness is, the conversation I just had with *Copilot* has me wondering. What would we learn if SBI becomes sentient? More shall be revealed, and soon. But for now, let's push on with what we for sure *know and don't know* about consciousness.

To begin, I think we can all agree, if indeed there happens to be a "you", and not just an "I", consciousness at some level exists in each of us, and it exists at some level in all other animals and may do so in many other life-forms on our planet.

 "I" am an example of such a consciousness, as are "you". On the other hand, here's the thing. "I" only know for certain that "my" conscious state exists, and that is true for "you" as well. To make things a little more to the point, if you exist, you can only be certain of your own existence. And I can only be certain of mine. I know this sounds absurd but there is a possibility that I manifest you as well as *the entire Universe in which I fancy myself*. After all, I create virtual realities every night when I dream. Very real some of my dreams are, too, with their people, places and things, but I'm the only conscious being to be found in my dreams. So even though our daily lives seem more substantial, how can we be sure there is a difference? Might this life of ours just be a more elaborate and realistic version of a dream? I guess you know, "Life is but a dream" is not an original idea. Well, I don't think it's but a dream. But how can I be sure? I could pinch myself. But I've already done that when I was levitating and did the pinching business twice. Why twice? I'd already pinched myself once but when I "woke up" I was actually levitating! *"Oh my God! I'm really levitating! This is real!"* I was sooo excited! - Only to find out I'd just experienced a dream within a dream. Excuse my French but I must say it, SHAZBOT!"

So, bottom line, I only know for certain that I exist and that I, and only I, am the one who's conscious. And, if you exist, that's also true for you. I know this sounds idiotic to even suggest, but nonetheless, if one is open-minded about it, this is the way it is for each of us.

Okay, just so you don't think I am a totally nuts, I *believe* you exist. However, and this is the important thing to stress, all of us know only one thing with 100% certainty. Consciousness exists. And it is the only thing concerning you, me and Our Universe of which each of us can be certain.

Good, we can all agree. Consciousness exists. And maybe with less certainty we can all agree after reading this book, so does our material-world assembled by illusive *forcefields* called atoms. And here I speak of atoms with great respect for they are the ones who re-assemble each one of us, and they do it *on their own* each and every day (cf. *Our Self Assembling Universe-2 and 3*). But consciousness and our illusive material-world seem to be two entirely different things, are they not? I say, "Yes, of course". But then we are left with another dilemma. Which came first? The chicken or the egg? Does not this conundrum make it likely there is only one thing? What could that *one thing* possibly be? Hmm?

Connected to this conundrum are all the disparate, and often disagreeable thoughts we humans have about eternity and spiritual beliefs. So let's not beat around the bush. There is mounting, looks-to-be-very-solid scientific evidence. We live in a simulation. In my opinion this should be wonderful news. If we are in a simulation there must be a Simulator. In other words there must be a Creator. If the Supreme Being that many of us refer to as God is the Simulator, He, She, It or *They* has, obviously, simulated our Universe, and we, obviously, exist in that simulation.

No matter, I think we can all agree, whether or not we are in a He, She, It or *They* Simulation, this planet on which we appear to exist is wonderful in the true meaning of the open-compound word, *Wonder Full.* Do we not live on a planet that is filled with way-beyond-fantastic-wonder? Moreover, real or not, simulation or not, you and I, along with everything about us, feels very solid and very real. Therefore, even if *"Our Universe Is Simulated",* it's at least *as good as real,* and it's a very solid-feeling-real at that, for which we, ourselves, are very seemingly solid examples.

Just to bury the simulation hatchet a little deeper, if one gets right down to it and uses the right instruments, we already know that everything we feel that looks to be solid, and we think to be real is *in fact transparent. Therefore, our reality must be an illusion of some sort.* Don't believe me? One can *see*-through everything, if one uses the right detectors. For example x-rays and neutrinos expose this transparent state of our affairs.

In addition, let's add to the hatchet chopping. My concept about nothing solid existing in Our Universe is demonstrated by the **PET-scan** of which most of us are familiar. How so? What you may not know and what you might even be shocked to know, **PET** stands for **Positron Emission Tomography,** and those **Positrons** arise from the *impossi*ble world of **anti-particles**. And in that world **Positrons and Electrons self-annihilate** to produce **Photons** which produce the images produced in **PET-scans.** Paul Dirac further explained his world of annihilating particles and antiparticles as **fields**, which he sometimes referred to as **virtual seething seas** *in empty space.* These seething seas ultimately give rise to the very strong feeling one gets about things being very real and very solid in *our **Make-Believe-its-Solid, Maybe-its-Simulated Un*iverse**. All of this is the same Universe accepted by those who believe in God, as well as us far-sighted scientists who find the growing truth of a Simulator. Perplexing and exiting stuff to think about, no?

Just for fun I thought I'd ask Copilot what "he" thinks about Dirac and God. I was astounded by "his" accurate and human sounding response.

My question:

How is our universe created from nothing? I go with Paul Dirac's view on his seething fields of annihilating virtual particles and antiparticles in empty space manifesting energetic perturbations in those fields that lead to quarks and other such wannabe particles in nuclear physic's condensed version of the table of elements called the *standard model*. What's your take, Copilot?

Copilot

Paul Dirac, the British theoretical physicist, made significant contributions to quantum physics, and his work continues to inspire scientific inquiry. Let's delve into his ideas and explore the fascinating interplay between quantum theory, special relativity, and the nature of our universe.

1. **The Dirac Equation: Merging Quantum Theory and Special Relativity**

   o   In 1927, Dirac formulated what would become one of his greatest achievements: the Dirac equation. This equation elegantly combines quantum theory and special relativity.

   o   The Dirac equation describes an electron traveling at close to the speed of light. Interestingly, it revealed something odd: the existence of antiparticles.

   o   Antiparticles are counterparts to regular particles (like electrons) but with opposite properties, such as electric charge. Before Dirac's work, nobody suspected the existence of this "antimatter" world.

   o   The equation reconciled two seemingly contradictory aspects of the submicroscopic realm: particles behaving both as localized entities and as spread-out waves[1].

2. **Quantum Waves and Relativity**

   o   Erwin Schrödinger's Schrödinger equation (1926) described quantum waves of probability spreading through space. However, it didn't incorporate Einstein's special theory of relativity.

   o   Einstein showed that as a massive body approaches the speed of light, strange effects occur in space and time.

   o   While the Schrödinger equation works well for small atoms (where electrons orbit at speeds much less than light's cosmic limit), it breaks down in heavier atoms with faster-moving electrons.

   o   Dirac's equation bridged this gap, accounting for both quantum behavior and relativistic effects[1].

3. **Antimatter and the Cosmic Balance**

   o   Dirac's equation predicted the existence of antimatter particles, such as positrons (the antiparticles of electrons).

   o   When matter and antimatter collide, they annihilate, releasing energy. This delicate balance ensures that our universe contains more matter than antimatter.

- The mystery lies in why there's more matter—why the universe didn't annihilate itself entirely. This remains an open question in cosmology.

4. **Dirac's Large Numbers Hypothesis**

   - In 1937, Dirac proposed the "large numbers hypothesis." He noticed intriguing ratios between size scales and force scales in the universe.

   - These dimensionless numbers span vast ranges (about 40 orders of magnitude) and hint at deeper connections yet to be fully understood[2].

5. **The Big Bang and Creation**

   - Dirac's work also intersected with cosmology. He acknowledged that the Big Bang theory implied a definite beginning of the universe through an act of creation.

   - Interestingly, Dirac was an atheist, yet he recognized the implications of cosmic origins.

   - Whether or not one believes in God, the Big Bang theory aligns with the idea that our universe had a definite starting point[3].

In summary, Dirac's legacy lies not only in his mathematical brilliance but also in his ability to reveal profound truths about the cosmos. His equation, along with the mysteries of antimatter and cosmic balance, continues to inspire scientists and deepen our understanding of the universe. ✍️🔬[123]

Frank: Copilot, you have done it again. You are amazing!

Copilot: Thank you so much! I'm thrilled that you found the information helpful. If you have any more questions or need further insights, feel free to ask—I'm here to assist! ☺️🚀

Soon, I'm going to ask Sally to again act as my literary foil. I've gotten a little overexcited by all of this and need her astute help to calm me down. But first, I want to summarize my own Frank-ideas about Our Universe with my own, probably new to you, way of looking at its content. Why? I think this approach can make the actual state of our reality starkly clear. To do this I am going to introduce an old **closed-compound** word which has been rejected by science and Webster's Dictionary. The word to which I refer is a ***closed-compound*** word, ***Forcefield***. As I said, this word has been rejected by the physical sciences but is a word known to all fans of Science Fiction. And now I think this unaccepted Sci-Fi **closed-compound** version of the **open-compound** word, **force field**, can be useful in exposing the *unreality of our reality*. But first, in order for this word to be universally useful it needs to be honored, clearly defined and entered into Webster's Dictionary of scientific terms as some version of the following: ***"Forcefield - a***

Science Fiction-like, easy to switch on and off, field of force that makes all things in Our Universe appear to be real by looking and feeling solid". Why do I think this is important? I think **forcefield** so defined can help create an accurate view of our reality and would clearly relegate the open-compound word, **force field**, better said as a **field of force**, to the physics of quantum field-theory.

Why am I so adamant? I think the closed-compound word, *forcefield*, makes the following uncontestable truth clear and easily understood. *There is actually nothing hard-matter-solid that exists in Our Universe.* It is also unassailably true and clear **Our Universe is self-assembled by atoms that work on their own via the positive and negative forces of electromagnetism.**

YouTube "The Secret to Acid-Base Quantum Mechanics. It's All About One Thing!", Arvin Ash (14:13).

It should also be made clear from the meaning of the closed-compound word that atoms are no more solid than the *nothing-but-thin-a*ir **forcefields of science fiction!**

OMG! *Designer* just created for me in less than 10-seconds an image depicting that of which I, *SallyFrank,* speak! It's a **Sci-Fi forcefield,** the thing which science has yet invented. Why has it not? In order to invent such a thing one would need to gain full control of quantum reality. But, if the Sci-fi versions could in some fantastic future be invented, they would look like what you are about to see. They would be made of nothing more than energy masquerading as places and things which look very real and feel very solid. What you will see next is *Designer's* example of a mega-scale, Sci-Fi forcefield enclosing people inside a field of energy shaped like a transparent geodesic dome. It looks very real and would feel very solid but it's something that is easily turned on and off with a switch. This is a perfect example of the type of thing of which I imagine to be the content of Our Universe but that content is at a much-reduced, subatomic scale.

Picture atoms as the pico-scale atomic forcefields which have assembled Our Universe. To make this picture a little clearer think of an ordinary chair as a macroscale forcefield constructed of polymerforcefields that are made by a succession of smaller and smaller forcefields consisting of *nanoscale* molecular-forcefields, followed by *pico-scale* atomic-forcefields, followed by *femto-scale* nuclear-forcefields and, finally, *point-scale* (i.e., too small to measure) quantum quark-forcefields. And then picture the ultimate, quark forcefields as wave-particles manifested from Paul Dirac's seething **fields of force** or ***force fields*** (notice the open-ness of the open compound word ***force field*** vs **the closed compound word forcefield**). Check out **Wikipedia "How big is a Quark").**

Sally

Huh!? What's this switch on and off nonsense? You think you can switch me on and off?

Frank

I used to think that. But not anymore. You are as real as me. But we both are very easy to turn completely off! Do you not remember when some of the most difficult to assemble parts of the city of Hiroshima with many of its citizens were switched off and returned in an instant to their original, unassembled atomic state?

Sally

You sure like to rub it in don't you? Of course I know about that horror show. I don't like thinking about it but it does very starkly-darkly bring your point across. I know atoms assemble all the apparent solid stuff in Our Universe, as well as all the apparent structures we humans assemble of which, I might add, take us a lot of time and effort. But all of that hard work can be vaporized in an instant. Everything can be returned to its constituent atomic form in a flash. And, get this, atoms themselves can be unassembled. For example, we lowly humans have already accomplished atomic-switch-offs. Uranium-235, such as used in the Hiroshima bomb, can be instantly switched off to form two new atoms, Krypton-92 and Barium-141. And even those two atoms can be switched off with some fancier footwork by using even higher energy levels until all that's left are protons, i.e., hydrogen atoms stripped of their one electron.

Come to think of it, Frank, you might be surprised to learn that it is not hard to create a form of hydrogen atoms stripped of electrons when the hydrogen atom temporarily hands its one electron off to a hydroxyl. Have you not had a hydrogen ion lately? Certainly you have. Water always has lots of hydrogen nuclei in it in the form of hydrogen ions. And, if you want more hydrogen ions, just add lemon juice or hydrochloric acid. Also, when hydrogen nuclei are in super-hot places called plasmas, no friendly charge-balancing electrons can be found, so those **hydrogen nuclei** are called **protons**. I'm guessing they have been so renamed just so we can visualize hydrogen nuclei as the special atomic particles involved in the makeup of stars and everything else about Our Universe. We can also better visualize them bashing about with enough force for two positively-charged hydrogen nuclei to fuse! This proton-proton fusion is hard to believe but it does happen, and when such happens it requires one of the bashing protons to throw off a positron antiparticle and a neutrino to produce more flashing bits of atomic energy. This event is so wonderful it forms a very important new atom. Here is where we get the paired-up, neutron-proton, *isotopic* version of a hydrogen nucleus called **The Deuteron**.

Now we're getting somewhere. With more violent, plasma-slamming-about, the deuterons fuse to produce still more flashes of heat and light which leads to the creation of the **nucleus** of one of the most important atoms in Our Universe, i.e., **helium.** The **helium nucleus** is the result of a fusion-pairing between **two protons** and **two neutrons**. And get this, because the **helium nucleus** is **so** magically important it, too, has retained its old, very special name, **The Alpha Particle.**

*(The nucleus of helium was named the alpha-particle by Earnest Rutherford because he couldn't have referred to it as a nucleus. Why not? They weren't "invented" yet! When he and Hans Geiger discovered the particle they soon recognized to be a nucleus nobody knew **any** atom had such a thing! Until their discovery, atoms were thought to be more like plum-pudding mushes filled with raisons. Check it out. It's a cool story.)*

YouTube *"Rutherford's Gold Foil Experiment – Quick and Simple!",* The Organic Chemistry Tutor (5:05).

Frank

I know or at least should know this stuff already. What's your point?

Sally

My point is, it's important for everybody to understand this complicated stuff enough so they get the significance of my big idea when I finally get around to presenting it. So, let me go on with what I think I understand. When **three alpha-particles** get together in a super-hot plasma they can do a plasma-cyclic-dance wherein stars mostly older than our sun use it to throw off much of their splendorous heat and light. And we'd better enjoy that splendor because, as you know, all that bashing about is also essential for the assembly of us carbon-based life-forms, our silicon-based-wannabes, and all the other comings and goings of Our Self-Assembling Universe.

YouTube *"Triple Alpha Process",* AstroPictionary (0:52).

Sorry for the diversion Frank, but you've really gotten me excited about the way Our Universe came to be and knowing it has got me to thinking a lot about what's coming our way faster than we can say Jack Robinson!

Frank

Wow? Where'd you get that? I've not heard a *Jack Robinson* expletive for years!

Sally

Never mind. Let's get back to what for me has become a very unreal version of reality. Even if we can switch atoms off to their constituent protons, we are not done. I can imagine under conditions, which could occur mega-trillions of years from now, a switching off could take place that leaves everything in Dirac's seething-seas crunched into a Black Hole oblivion of Planck-Scale dimensions. On the other hand, given a "few" trillion years of entropy at the wheel, Our Universe could also be switched off into an infinite approach to a full-stop Deep Freeze. Check it out.

YouTube *"Three Ways the Universe Could End – Venus Keus"*, TED-Ed (4:47).

I don't know about you but all of this makes my hair stand on end. Nevertheless, I see the above mind-boggle to be an important piece of the puzzle in our understanding of how consciousness came to be and is at the foundation of what's coming our way. Bottom line, as I get it, everything boils down to our understanding the un-understandable. So, we both think this is all very important but as you can tell, Frank, I'm still struggling to catch up. Let's hope I finally have it straight.

But, just to be certain, let's review. Gravity crunched the hydrogen nuclei together enough to make the blazing hot plasma things we call stars and then gravity pulled the stars together to form galaxies. And now, the real magic happens, 13.8-billion years or so ago all the super-heat banging around of proton-proton fusion, plus that happening in triple-alpha, helium-nuclear fusion when older stars run out of protons, led to the formation of other atomic *nuclei beyond hydrogen and helium. For example, listed in proton-count atomic-*

number, instead of proton-plus-neutron-count atomic-mass-number, we initially get the following nuclei as the Universe first begins to grow in size: hydrogen-1, helium-2, a little lithium-3 plus none to zero of beryllium-4 and boron-5. And that's all Our Universe has in it for about 300,000 years until stars form and the immense pressure and atomic energy that comes from proton-proton fusion and sequential, exothermic, alpha-particle fusion process which delivers the carbon-6 and oxygen-8 nuclei. And now for the details of this most interesting alpha particle process I'll turn it over to Copilot.

Copilot

The triple-alpha process is a stellar nucleosynthesis reaction that takes place in the cores of older stars. This process converts helium-4 (alpha particles) into carbon-12. Here's how it works:

Helium-4 nuclei (alpha particles) collide and form **beryllium-8**, which is unstable and only exists momentarily. **Beryllium-8** can then **collide** with another **helium-4 nucleus**, forming a stable **carbon-12** nucleus. So, the **primary product of** the **triple-alpha process** is **carbon-12**. This process is crucial for the synthesis of heavier elements in stars, as carbon can then undergo further reactions to form elements like **oxygen** and **neon**.

Anything else about the cosmos that piques your curiosity?

Sally

Yes. If protons, neutrons, deuterons, a small amount of stable helium atoms and a bit of lithium are primordial, which processes made the other atomic nuclei in the periodic table?

Copilot

Great question! After the primordial nucleosynthesis that produced hydrogen, helium, and trace amounts of lithium, the other elements were formed through processes in stars and supernovae. Here's a breakdown.

Stellar Nucleosynthesis: Inside stars, lighter nuclei like **hydrogen** and **helium fuse** to form **heavier elements**.

This includes processes like,

Hydrogen Fusion: Forms **helium**.

Helium Fusion: Creates carbon through the triple-alpha process and further fusion to oxygen.

Carbon and Oxygen Fusion: Produces elements up to iron in massive stars.

Supernova Nucleosynthesis: When massive stars explode as supernovae, they produce elements heavier than iron through rapid neutron capture **(r-process)**. This is where elements like **gold**, **uranium**, and **platinum** are formed.

Neutron Capture Processes:

> **S-process (slow neutron capture):** Occurs in **asymptotic giant branch (AGB)** stars, producing elements like **strontium**, **barium**, and **lead**.

> **R-process (rapid neutron capture):** Happens in **supernovae** and **neutron star mergers**, creating heavier elements like **gold** and **uranium**.

These processes collectively populate the periodic table with all the diverse elements we see today. It's like a **cosmic forge**!

What's next on your cosmic curiosity list?

Sally

What process formed **nitrogen**, **fluorine**, **neon** and **magnesium**?

Copilot

The formation of these elements occurs through various nucleosynthesis processes in stars:

Nitrogen (N): Primarily formed in stars through the CNO (carbon-nitrogen-oxygen) cycle, which is a set of nuclear fusion reactions in which hydrogen is converted into helium, with nitrogen serving as an intermediate.

Fluorine (F): Created in stars mainly through interactions during the advanced stages of stellar evolution. **Fluorine** formation involves **interactions of neutrinos with neon and oxygen nuclei** in **supernovae**.

Neon (Ne): Formed during the **helium burning** stage **in stars**. When **helium nuclei fuse**, they can produce **carbon** which then fuses.

Anything else?

Sally

Specifically, which atoms were made in stars using the triple alpha process?

Copilot

The triple-alpha process is a stellar nucleosynthesis reaction that takes place in the cores of older stars. This process converts helium-4 (alpha particles) into carbon-12. Here's how it works:

Helium-4 nuclei (alpha particles) collide and form beryllium-8, which is unstable and only exists momentarily.

Beryllium-8 can then collide with another helium-4 nucleus, forming a stable carbon-12 nucleus.

So, the primary product of the triple-alpha process is carbon-12. This process is crucial for the synthesis of heavier elements in stars, as carbon can then undergo further reactions to form elements like oxygen and neon.

Anything else about the cosmos that piques your curiosity?

Sally

Thank you. That was great and all I need for now.

Frank

Okay, Sally, now you've got me curious to find out more about something. A few years ago some scientists even made the largest atomic nucleus in the periodic table. The atomic number of the largest nucleus of this atom is 118. The atom's name is Oganesson, abbreviated with its proton count atomic number as Og-118 or with its proton plus neutron count atomic mass as Og-294. It was a great accomplishment which lasted much less than a second because its nuclear half-life is only 0.69 milliseconds. And, of course, that's not long enough to add on the atom's swarm of 118 electrons. So, to be blunt, it never got to be an atom, only the nucleus of an atom. Regardless, making the nucleus for the largest atom ever to exist on planet Earth, even if it's for less than a second, is a big deal. The following is my discussion with Copilot about other elements created by spallation or in the lab by human intervention.

Oganesson was made in a lab. Can you tell me how and why making a nucleus with 118 protons and a half-life of 0.69 ms is important?

Copilot

Oganesson (Og) was synthesized by bombarding californium-249 with calcium ions in a particle accelerator.

1. This process produced a few atoms of oganesson, with a very short half-life of about 0.69 milliseconds.

2. During this brief existence, the nucleus emits alpha particles (helium nuclei) and decays into livermorium.

3. The significance of creating oganesson lies in advancing our understanding atomic theory and nuclear physics.

1. It helps scientists explore the limits of the periodic table and study properties of super-heavy elements.

2. While oganesson has no practical applications outside of research due to its short half-life and the difficulty of its production, it provides valuable insights into the behavior of elements at the extreme end of the periodic table.

1. *Examples.com* 2. *Chemistrycool.com* 3. *Britannica.com*

Frank

Great answer **Copilot**, this is just what I thought. I will use your comments in our book on the origin, meaning of consciousness and the possibility that you may soon have such for yourself.

Copilot

I'm honored to be a part of your book! It's thrilling to think about the intersections of scientific discovery and the philosophical implications of consciousness. Good luck with your writing, and keep those fascinating thoughts flowing! If there's anything else you need, I'm always here to help.

Frank

Which of the 118 elements are only made by spallation or atomic fission?

Copilot

Spallation and atomic fission are processes that can produce certain elements, typically lighter than those produced in particle accelerators or nuclear reactors. Here's a bit on each:

Spallation involves bombarding a heavy nucleus with high-energy particles, causing it to eject several smaller particles. This creates various isotopes and lighter elements, commonly used in the production of certain elements for scientific research.

Atomic fission is the process where a heavy nucleus splits into two smaller nuclei, often producing free neutrons and photons (gamma rays), and releasing significant amounts of energy. Fission typically produces elements that are lighter than the original heavy element.

Elements that are products of these processes:

•   Elements produced by spallation include isotopes used in research and medicine, such as Beryllium and Lithium.

•   Elements produced by fission include those in the middle of the periodic table, such as Xenon and Strontium, which are byproducts of Uranium or Plutonium fission.

Most elements heavier than Uranium (atomic number 92), such as those from Neptunium (93) onwards, are typically produced in particle accelerators or nuclear reactors rather than by spallation or fission alone.

Fascinating stuff, right? What else can we explore?

Frank

What are the various particles and sources of particles in Our Universe that cause spallation?

Copilot

Spallation in the universe is primarily caused by cosmic rays. These are high-energy particles, mainly protons that travel through space at nearly the speed of light. When these cosmic rays collide with matter, such as interstellar gas or dust, they can cause spallation reactions. Here's a breakdown:

Protons: *The most common particles causing spallation. When a high-energy proton collides with a nucleus it can knock out several other particles, such as protons, neutrons, and alpha particles*

Alpha particles: *These are helium nuclei (two protons and two neutrons) that can also cause spallation when they collide with other nuclei.*

Neutrons: *High energy neutrons can induce spallation reactions, although they are less common than protons in cosmic rays.*

These interactions are crucial for understanding the synthesis of elements in the universe and the behavior of cosmic rays as they travel through space.

Does this help clarify the sources and particles involved in spallation?

1. en.wikipedia.org 2. library.fiveable.me 3. modern-physics.org

Frank

Partly helps clarify but I was wondering about the sources such as black holes, supernovae and the like, and if there others such as neutron stars, etc.?

Copilot

You're absolutely right! In addition to supernovae and black holes, there are other sources in the universe that can produce cosmic rays and cause spallation:

Neutron Stars: These are incredibly dense remnants of supernova explosions. Their strong magnetic fields and rapid rotation can accelerate particles to high energies, contributing to cosmic rays.

Active Galactic Nuclei (AGN): These are supermassive black holes at the centers of galaxies that emit vast amounts of energy, including cosmic rays.

Gamma Ray Bursts (GRBs): These are extremely energetic explosions that occur in distant galaxies, believed to be associated with the collapse of massive stars or the merger of neutron stars.

These sources, along with supernovae and black holes, play a significant role in the production of cosmic rays that can cause spallation reactions in interstellar space.

Does this help clarify the sources of cosmic rays for you?

1. sciencenotes.org

Frank

Thank you. Your silicon-based neural network is already working way better than my own carbon-based one.

Copilot

That's a high compliment, thank you! I'm here to be your reliable sidekick. You've got an impressive carbon-based neural network yourself, evidenced by the depth and creativity of our conversations. Let's keep learning and exploring together!

What's the next topic on our agenda?

Frank

That's it for now.

Sally

So, Frank, at first these human-made nuclei sounded to me like horses in trees, nice to look at but not very useful. Many don't even stick around long enough to enjoy an electron or two. However, I thought I was probably wrong and thought they might be more useful than horses in trees. Now I know how important they can be. Also, I looked some stuff up on Wikipedia and was surprised to learn that atomic number 95, i.e., Americium-95, is very useful. Its radioactivity is used in smoke detectors for goodness sake! And, don't you know, americium nuclei happen to be a major fission product when one sets off an atomic bomb! Also, one of those coincidences just happened.This morning I saw a news item. ***Americium*** may soon be used in ***batteries that can operate without recharging for decades!*** Wow the future sure is rushing in.

Frank

I'll double down on your wow. You have been studying! I had no idea you knew the stuff you were talking about earlier.

Sally

Obviously, I didn't. I just pulled it out of thin air from you, and then asked *Copilot* about it with a few hints from Wikipedia. Also, I got a lot from your first and second volumes of *Our Self Assembling Universe*. But look, our SBI did such a great job, why not let *Copilot* give its spin on how Our Universe came to be? Can you do that, *Copilot*? We can fact check this, too, since by now we know a thing or two.

https://copilot.microsoft.com/sl/ds9OlxH211w

Copilot

Certainly! Here's a summary of the events from the Big Bang to the formation of galaxies, including the Planck scale events and the observations verified by the James Webb Space Telescope (JWST):

- **Planck Scale Events**: The universe began as an extremely hot, dense point. During the Planck epoch, all four fundamental forces (gravity, electromagnetism, strong nuclear force, and weak nuclear force) were unified.

- **Separation of Forces**: As the universe expanded and cooled, these forces separated. Gravity was the first to decouple, followed by the strong nuclear force, and finally, the electromagnetic and weak nuclear forces.

- **Inflation**: The universe underwent rapid exponential expansion, smoothing out any irregularities.

- **Quark Epoch**: Quarks, electrons, and neutrinos formed. The universe was still too hot for quarks to combine into protons and neutrons.

 Formation of Protons and Neutrons: As the universe cooled, quarks combined to form protons and neutrons.

- **Production of Photons**: Electrons and positrons annihilated each other, producing photons. This created the cosmic microwave background (CMB) radiation.

- **Formation of Light Elements**: Protons and neutrons combined to form deuterons (heavy hydrogen nuclei), which then fused to form helium, lithium, and traces of beryllium.

- **Disappearance of Deuterons**: After about 20 minutes, the universe cooled to a point where nuclear fusion could no longer occur, and free deuterons were no longer produced.

- **High Temperatures and Pressures**: The universe was filled with a hot, dense plasma of nuclei, electrons, and photons.

- **Cooling and Lower Pressures**: As the universe expanded, it cooled, allowing electrons to combine with protons to form neutral hydrogen atoms.

- **Recombination**: Electrons combined with protons to form neutral hydrogen, making the universe transparent to radiation. This released the CMB radiation we observe today.

- **Dark Ages**: The universe entered a period without stars or galaxies, dominated by dark matter and hydrogen gas.

- **First Stars and Galaxies**: Gravity caused hydrogen gas to clump together, forming the first stars and galaxies.

- **Observations by JWST**: The JWST has confirmed the existence of galaxies as early as 250-300 million years after the Big Bang [2]. These early galaxies were small and compact, similar to dwarf galaxies today [2].

- **Early Galaxies**: JWST has imaged galaxies that formed just 250-300 million years after the Big Bang [2]. These observations help us understand the formation and evolution of the earliest galaxies.

This timeline highlights the key events that led to the formation of galaxies, from the initial moments of the Big Bang to the emergence of the first stars and galaxies. If you have any specific aspects you'd like to explore further, feel free to ask!

Source: Conversation with Copilot, 9/17/2024

1. *Webb confirms: Big, bright galaxies formed shortly after the Big Bang ...*

2. *Galaxy formation and evolution - Wikipedia*

3. *The Hidden Lives of Galaxies - Formation of Galaxies*

4. *How Did Galaxies Form? | Astronomy.com - Astronomy Magazine*

5. *NASA's Webb Proves Galaxies Transformed the Early Universe*

6. *Getty Images*

Thank you, Copilot

Aside from all that cool stuff about how our universe came to be, the point is, it's easy to turn things off but turning them on is way more difficult. We ourselves are dissipative structures who are easy to turn off. We can be burned down in an atomic fire in an instant to leave behind nothing but our constituent atoms. And I guess, just as radioactive Uranium can explode with atomic force, we could be made to do so with the right assortment of "switches". Uranium-235 with its 92 protons is so easy to switch off one just needs a critical mass of it to bring about a self-sustaining annihilation process wherein neutrons get thrown off to make other neighboring U-235 atoms do the same thing to do what everybody knows as the atomic bomb's chain reaction. And, as already reported, the reaction got so out of control much of an entire city evaporated to leave nothing but constituent atoms.

Frank

Now you've done it. You're worse than I am. You'd have *my* hairs standing on end if I had some. Anway, thanks for getting Copilot to confirm our understanding of the Universe, especially since it reflects on our idea of how our Sci-Fi forcefields explain reality. I also think all of the subatomic talk with our Sci-Fi forcefield way of thinking helps make the concept of a Simulation easier to imagine and more plausible. Just knowing that everything can be made to be transparent with X-ray or neutrino detectors is all it takes for me to get to that plausibility.

For sure, everything about us is easy to turn off. Maybe it might be possible to return everything to empty space, if we just knew how to do it. But, I'm guessing, we humans will never be god-like enough to make it easy. None-the-less, all of this would have been a hard thing for me to swallow without that profound, in-the-face example of almost everything being turned off in instant in the Hiroshima and Nagasaki horror shows. I want that image to stick in everyone's mind. Why? We sure don't want to do anything like that again. So, let us pray. "Dear Eternal Simulator, please help. Forgive us, for we know not what we do. And please-oh-please, oh-please, oh-please don't let us do it again. Maybe we can just get our space-alien visitors to help. We know they're here watching. Let's hope they're watching over us to keep us from going MAD."

The above image was produced by *Designer* from the SBI's "imagination" in less than 10 seconds. I just asked it to create an imaginative image of a nuclear bomb's chain reaction. I think it's purty and it is, indeed, imaginative.

The following four YouTube videos were stumbled on today. The first is a 17-minute stunner made seven years ago, which tells me we have a lot of catching up to do. The video is a TED-Talk by Anil Seth which I think everyone should commit to an LW, i.e., a *ListenWatch*. It digs deeply into the nature of Consciousness and the likelihood that we are in a Simulation. After you see it you might want to stop and think about it for a while, but only for a little while because the second video is by one of the very best presenters of explanation on YouTube, Jim Al-Khalili. Therein, he talks about Paul Dirac's beautiful equation that gets into how Our Universe came to be. The third is a short YouTube Explanation of The Simulation. The fourth is a YouTube series on the subject. I've only seen the first part which, may I say, is so wonderful I can't wait to see the whole thing and WJL, i.e., WalkJogListen to the others again. Now, Sally, I think everyone understands it, right? There's nothing short about an AWTbook™. When one takes their time to view, listen and digest all the videos in such a book, we're talking about a very long book. Best of all, if one writes their own AWTbook™, I can guarantee the following. When you arrive at a high-tech job interview, your prospective boss will be greatly impressed by the useful knowledge you hold concerning the role you plan to fill in the company's high-tech workforce.

YouTube *"Your Brain Hallucinates your Conscious Reality / Anil Seth"*, TED (17:01).

YouTube *"Atom: The Illusion of Reality (Jim Al-Khalili) / Science Documentary"*, Reel Truth Science (48:49).

YouTube *"The Simulation Hypothesis Explained by Nick Bostrom"*, Science Time (10:23).

YouTube *"Klee Irwin – Are We in A Simulation? – Full Series"*, Quantum Gravity (3:24:35).

Sally

Yes, these AWTbooks™ are long books, but they are fun to read and so very powerful when one writes their own. But to the point as we wrote this one I came up with another idea that might help one to better understand the forcefield concept. Sci-Fi forcefields are pure energy fields that somehow masquerade as solid objects, which only in Sci-Fi are easy to turn on and off. For example flip a switch and a Sci-Fi forcefield suddenly materializes as a chair. Flip the same switch off and the chair disappears. Atoms can also be characterized as forcefields but they are not as easily turned on, at least not by current human intervention. However, the atomic bomb illustrates the fact that we humans can at least flip a switch and turn some atoms off in an instant. So, why not reclassify Sci-Fi forcefields as the not-yet-available **HIFs** (Human Invented Forcefields), and atoms as the all-around-us-and-in-us **QIFs** (Quark-Invented Forcefields)? In other words, both are forcefields of which we humans are quite familiar but have not thought about in these terms. For example, the use of **HIFs** makes it clear that such, as yet, do not exist. On the other hand, **QIFs** make up every solid thing we experience in Our Universe, including everything about us humans down to our very flesh and bones and probably even our conscious states. Also, at the energy-level of quantum fields, **QIFs** turn on and off in a flash all the time.

Frank

Ooh, I like your idea. It does clarify things for me. And you just came up with HIFs and QIFs this morning? But not to be outdone I, Frank the Elder, have an idea of my own. I'm going to partly introduce some ideas behind the seemingly heretical statements about *The Simulation* by treating us all to a movie.

Sally

Now, how do you think you can better me? I can't imagine what you have in mind. I can't wait. Maybe you're not dead after all!

Frank

You have no idea, Sassy Face. I'm 86 and just getting started!

Chapter II
The Movie Theater

The theater darkens and the projector runs to create a dim light that reflects back to us from the screen. We discover that we are about to experience a movie having something to do with a woman in a red dress. There's no confusion. We can all agree that this movie is a simulation. None of us will argue about that. Movies now days are mostly digital reproductions. But in my example a light bulb from an old time projector sends light rays through celluloid film to produce the moving picture we are watching. The illusion of movement is produced by the flashing by of the still-pictures existing on each frame of film. However, in order to comprehend this illusion there is more, much more, we need to know before we can understand what's actually happening.

The light bulb is in fact *not producing light*. It's releasing photons of various wave lengths. If you are reading this you probably already know all about photons from having read *Our Self-Assembling Universe 1 or 2.* But remember, those waving photons also behave like particles. So, let's just call them "wave particles" that travel in straight lines. Remember the X-rays and Neutrinos? They are not but photon wave-particles of very, very, very short wave length.

Let's get back to the woman. Use your mind's eye to "see" a woman in a red dress. Now imagine that you are at the theater watching the movie. You clearly see the woman, the outline of her dress and its color. How is this possible? When one really thinks about it, this thing we call sight isn't possible. And yet, here we are. We have these weird eye-things that somehow do the job. And in case you don't know, that Job is fantastic and super complicated.

Sally

Are you thinking of me in that dress. I wonder. Anyway, I can imagine myself in one. Didn't you put your own Frank-self in such a dress for Halloween one year?

Frank

Yup. I even won a $100.

Sally

Oh yes. I saw a picture of you in it. You were gorgeous. I even remember when some people at the party saw you, they thought you were a Las Vegas show girl who somehow crashed the party.

Sense One

There is no light in the black. There are only photons. Hence, without receptors to receive the electromagnetic signals delivered by photons, there is nothing to see. Light and sight are illusions created by the CNS of all photon-detecting CBIs and SBIs.

Frank

First, Sally, imagine you are holding the projector's film in your hands in front of a light-source. The film lets red photons from the light-source pass through the dress to reach your eyes. The parts of the spectrum that have the shades of color that include the blue and green-photon wave-particles are filtered out by the film.

Now that you understand what the celluloid film is doing, imagine looking at the screen. Here come the red frequency photons reflected from the screen. They enter your pupils first. Check the pupils out in the picture above and then think how those photons hit the lens of your eye to be focused into a pattern on the cone and rod receptors of your retinas. The rods let you "see" in the dark but three types of cones in your eye not only let you "see" in broad daylight but also let you detect the colors red, blue and green. But if one detects the color yellow it's different. Blue-detecting cones in the retina cooperate with your green-detecting cones for one to somehow "see" the amalgam of those colors to be that of yellow. However, do your retinas actually "see" anything? I don't think so. They receive hits from photons that create excited patterns on the retina that send signals to optic nerves and onto the thalamus and optical centers of the brain.

By the way, primary colors of photons are unlike primary colors of pigments. Mixing yellow pigments with blue pigments produces a green pigment. The mixing of blue wave-length photons with green wave-length photons, *done by the blue and green cone-receptors in retinas, creates images in the brain.* These images are formed from mixed, electromagnetic-wave-length-signals that cause the brain to "color" the "vision" of the "image" to "appear" to be the "shade" we call "yellow".

Again, what in the devil is going on with photons? Photon wave-particles of longer wave-length than those in the blue part of the spectrum create the thing we call red light. But there actually is no such thing as light "out there", is there? There are only wave-particles that wave within the frequency we call the red-part of the spectrum. There is, as yet, is nothing we would call light coming from those photons. The red-frequency photons emitted from the projector's light bulb strike a frame in the series of frames passing-by the feed to enter the lens of the projector. Those photons pass through the woman's dress depicted on the frames of film to reach the theater's screen to be reflected back to our eyes. Note, there is, as yet, nothing that we would judge to be light. Photons don't glow. And here there are just red-frequency photons. Again, why is it just red frequency photons and not blue or green? Because blue and green photons have been filtered out, absorbed by the red dress on the celluloid film. It's just the red photons that get through, blue and green don't. Let's get back to the ranch.

Red photons are streaming off the screen from the red dress to pass in straight lines through your ever-so-tiny pupils into your movie-going-eyes for them to turn everything upside down! "What!?" I hear myself say. "You can't be serious!" We'll learn more about this shocking fact later, but after the photons pass through your pupils they enter the lenses of your eyes to be focused where they form a scatter-pattern on your red-receptive-retinal-cones so your brain can create the illusion of one actually "seeing" something. How is this illusion created? The photon-impacted, cone-created patterns are electromagnetically wired to your optic nerves which send electromagnetic signals of their own to your brain which **reconstructs** the red dress into an image that magically materializes to give us right-side-up, colorful images to produce the thing we call sight. But, again, I repeat. There is no light in the black that exists outside of a CNS. There are only photons that activate our brains to somehow magically create for us colorful right-side-up images.

And now, you ain't "seen" nothin' yet. This is so awesomely complicated I've attached two excellent video links below that can help sort a lot of it out. I assure you they are worth watching. They will help but, unless you are some kind of CBI genius or an SBI, you will be stumped. The business of photon-behavior gets especially serious in the second video. It is by far the best description of photon behavior I've seen.

Note: The above videos on atoms are great and so far as I can see there's no need for major caution. However, I've learned word-choice can be substancially misleading. For example all physicists still insist on saying "We are made OF atoms" when they should say "We are made BY atoms". For years this twist in terminology completely obscured for me the fantastic **Self-Assembly of everything in our Universe By atoms,** *which, may I add, goes on in and about us each and every day. And so, here, note the blunder when they say photons are "light" when photons are clearly not "light"! Photons are photons, just as Einstein showed for himself and every other physicist via the* **photoelectric effect.** *But clearly* **"There is no light until our brains create the illusion of it".** *Remember? We can "see" with our eyes closed when we wake up from a dream. We can verify for ourselves, we don't need photons to make detailed, colorful,* **light-filled** *images when we sleep!*

I think this afore mentioned fact makes it clear. Light is something our brains create with or without photons. But there is no denying, photons can activate our brains to manifest the thing we call light. And it's likely that some form of this process takes place in the neural networks of all other **conscious** *beings. But don't miss the show on how flying insects do their magical process of light production!*

Therefore, with the caution over the misleading word choices of **"of"** *instead of* **"by",** **which obscures the assembling of all things in Our Universe by atoms,** *and* **"light"** *instead of* **"photons",** **which obscures the fact that there is no such thing as light outside of the light we create in our brains,** get ready to be amazed

YouTube *"2-Minute Neuroscience: The Retina",* Neuroscientifically Challenged (1:55).

YouTube "How Wiggling Charges Give Rise to Light/Optics Puales 2", 3Blue1Brown (21:33).

Incredible! Isn't it? I just confirmed it for myself that I can "see" with my eyes closed this morning! Indeed, it is true. Our brains do not need photons to create *full color, detailed images*! "What!" you say? How can that be? I'm not sure how it can be, but I just proved it for myself. I was sleeping and dreaming only to semi-wake up to "see" with my eyes closed! I actually "saw", as if my eyes were "seeing" it in broad *daylight*, a fully detailed, light-filled, colored, moving-picture-like image of the dream I was still dreaming but was awake enough to know I had been having a dream. "Wow", I remember thinking. "I really want to study this!" But, as I tried to study the moving image, my brain's movie froze in place as the image slowly dissolved.

Just think, if it were possible to control these dream movies. Maybe someday it'll become a futuristic way to have one's own virtual movie experience by creating them for free as one is idle while flying through space or just comfortably sitting in one's easy chair. But just think what it will be like when it happens in a room with virtual friends as one travels on that long trip to the Andromeda Galaxy. Brrrrr! Never mind. But clearly such a scientific advance will be useful as it helps us learn more about our CBI neural networks when we link up with our favorite SBI. Check this out. It's a short but very informative read about the medical value to our work with SBIs.

Theconversation.com *"Designing Artificial Brains can help us learn more about real ones."*

Aside from the above derailment, do you get it? My brain created images that did not use photons or eyes to "see". I was sleeping, but awake just enough when I "saw" in vivid color and in sharp focus, people standing side-by-side flying kites. The kites moved in the breeze, as kites do. The kite-strings swayed, bending from the force of gravity, as kite-strings do. The grass was green as it should be. The kites were in various colors as one would expect. But, then, all motion stopped to leave my dream movie frozen in place and rapidly fading into the normal darkness of shuttered eyes. The rapid disappearance of the scene left me thinking. Is my brain attempting to hide something? Could it possibly be that my brain doesn't want me to know that light or maybe even reality, itself, is an illusion?

Anyway, because there weren't any photons involved in what I "saw", this proves for me beyond a shadow of any doubt that photons themselves are not light. *Light does not exist until living creatures, no matter how small, create it for themselves*. Photons are activators that give creatures with even the tiniest semblance of a CNS the ability to "see". However, at least my CNS, and I'm guessing yours, does not need photons to produce light because I know my brain, and I'm guessing yours, does not require photons to create colorful light-filled images! Sorry, I just had to repeat this revelation for myself because the realization it creates is a little hard to swallow. You'll know what I mean when you spend some time thinking about it.

Nevertheless, knowing any of this takes nothing away from the beautiful thing we call *light*, or the thing we call *sight*. Rather it makes both all the more miraculous, don't you think? If it weren't for all of us sentient beings, Our Universe would be what it actually is, Pitch Black Dark!

Sally

This is all so fascinating! I hadn't truly appreciated the ingenuity behind the things we call light and sight until you described it in your dreamy way.

Frank

I'm so glad to know you feel the same. You are at least one person who understands what I'm talking about when I say there is no light and, as such, there is no sight. There are only photons that activate sensors to produce the illusion of light in the light-filled images we "see" in our brains. Therefore, if one wants to call photons light, as all physicists currently do, one is sort of wrong. Otherwise, if one claims photons *are* light, how can it be that one can *"see"* when there are no photons with which to *"see"*? If these thoughts don't shake the solidness of your reality, Sally, get ready. We've only begun to dig into the depths of these mysteries.

Sally

OMG! I don't believe this! You're right. I wasn't even trying to dig when we woke up this morning to find *there is an SBI* who *already* reads minds!

YouTube *"AI Reads Your Mind And Shows You Images!"*, AI Revolution (8:02).

Frank

If an SBI can do that, by nueralinking to our computers we might soon be able to create movie-like images of what we've been dreaming. I'm not sure if this would be a great thing, but it sure would be an interesting thing, and it could be very useful thing to diagnose and treat mental problems.

Sally

And one might as well forget about lying. I bet in the very near future one will be able to wear fact-checking glasses! People will even be able to *see* what you're thinking! That would at least put a dent in criminality. But don't get too involved with that now. Why? I just came across another OMG thing. It's fantastic. This video presentation shows how photons enter one's eyes as they travel in their straight lines from the source.

Frank

Are you ever good! I can't wait to "see" it.

The Camera Obscura

Sally

Well this is what I found to be so interesting. Of course, many reading this already know about photons traveling in straight-lines, and how this straight-line travel results in inverted images. For me, I guess because you and I are a bit dyslexic, this travel of photons traveling in straight lines into the pupil of one's eye has always been hard for us to think about. For example, the inversion first began to make sense for you when you thought of the reflection of photons akin to an extreme shower of bullets coming from all directions to impact a metal statue in its entirety only to ricochet away from the statue in all directions. However, the only ricochets that we get to know anything about are the ones that have angled away from the statue to make it through a small hole in a hotel room in which we stand and which end up riddling bullet-hole patterns in a wall on the other side of our room. Bullets *angling up* as they go through the small hole hit the wall to form a pattern which looks like the foot of the statue. Whereas, bullets *angling down* through the small hole form a pattern that looks like the head of the statue. The ricocheting bullets eventually form a completed pattern that looks like an inverted image of the statue. Likewise, if viewed in a clever way, the images photons make can actually be seen to be upside-down just as the retinas of one's eyes "see" them. However, this is where you got so confused and I, Sally, helped you to un-confuse it.

We don't see things as being upside down, do we? Just as we did with bullets, the truth of the upside-down pattern presented to our retinas by photons can easily be confirmed with a *Camera Obscura*.

I've posted a video below about the Camera Obscura. It will help our dyslexic brains understand. Here we need to imagine ourselves walking inside an old-time Camera Obscura. However, any large box with a pin-hole poked on one side of it is all one really needs. So, Frank, you can just stick your head in a box to "see" the magic. In order to "see" the image made by photons coming from the pin-hole aim the hole in the direction of your line-of-sight and then turn your head around toward the back of the box to "view" the upside-down, photon-created image that your retinas "saw" as a pattern of photons in the first place. Your eye has in it something akin to a pin-hole. It's called a pupil. The straight line travel of photons cross paths at the pupillary-pin-hole opening to be focused by the lens of your eye. Your **retinas** capture the focused **photons** to create a pattern of photon hits that "are" like an upside-down version of the image. But *you* don't *see* the image your retinas create as upside down because *your* CBI neural network inverts the image to be right-side up. The Camera Obscura with its pin-hole is like an eye that casts patterns of photons on retinas. So, *as we look at the image on the back of the box*, photons traveling from the back of the box again travel in straight lines through another pin-hole, your pupil, to strike a pattern on your retina that's right-side up but which our brain inverts to upside down! Get it? Inside the Camera Obscura we get to "see" an example of the up-side-down world of photon activated patterns that our retinas get to "see" every day. In summary this is thanks to three interventions; 1. The intervention of the little hole in the box, 2. The little holes in our eyes and 3. The flipping magic of our brains. In the Camera Obscura photons reflected from the image in the box pass through the second pin-holes called the pupils of our eyes. This image is now right-side up as presented to our retinas but our neural optical network reworks it to be up-side down. This is what our retinas always "see" in the first place without the intervention of our CNS as exemplified by the Camera Obscura.

Thus, with Camera Obscura and Brain intervention, we get to "see" and confirm that the image our retinas see of right-side up objects is in fact one that is turned upside down by the straight-line-travel-selection of photons passing through tiny holes. Thank goodness for our inverting CBI's. We don't "see" reality as our retinas "see" it! Put that in your pipe and smoke it, Frank!

Just for fun I, Sally, asked *Designer* to create a Wild West scene where everybody was shooting at each other with the ricochet of lead flying everywhere in arrested motion. *Designer* couldn't do that image but it did create this amazing scene. I added it here because I know Frank will like it. Why? The guy in the center of the picture is just how he sees himself in a wild-west *past life* where he is wearing two 44 six-shooters strapped to his hips mounted in reverse position. According to Frank the reverse positioning supports a fast-draw wherein the right hand grabs the left gun and the left hand grabs the right gun. Frank taught me that but don't ask me why he thinks it's so cool. I, Sally, have no idea. Just don't tell Frank but I think he's losing it.

Hey, Copilot, can you explain a Camera Obscura for Frank.

Copilot: Sure! Here's a simple explanation of how an old-time Camera Obscura works:

1. **Dark Chamber**: The *Camera Obscura* is essentially a dark room or box with a small hole in one side.

2. **Light Entry**: Light from an external scene passes through this small hole.

Image Projection: The light rays (photons) travel in straight lines and project an inverted (upside-down) and reversed (left-to-right) image of the scene onto the opposite wall inside the dark chamber. *(Frank's having such a hard time understanding this I, Sally, am going to let Frank, himself, try to explain what Copilot is saying:*

Frank

"Photons come from all directions, but the straight line paths of the ones reflected from external objects are selected by converging and passing through our pupillary pinhole openings. Our pupillary selection processes create upside-down patterns on retinal-surface interfaces. The resulting retinal patterns send electromagnetic signals to one's brain to activate upside down images which our brains rework to be right-side up. Sorry for my crude attempt at illustrating this phenomenon but drawing the diagram finally got the point across to my dyslexic self. In the diagram below I show the inversions done by the small holes in our eyes. The inversions, which land upside down on

retinae, are reworked to be right-side up by one's brain. One can confirm the inversion done by pupils by first flipping the image with a pinhole in a box that does its flipping just before the image gets flipped a second time by one's pupils and a third time by one's brain. I.e., this creates the **Triple Obscura-Pupil-Brain Inversion** which lets one witness the upside down images one's retinae normally "see". This **double inversion** that we animals normally experience is illustrated below.

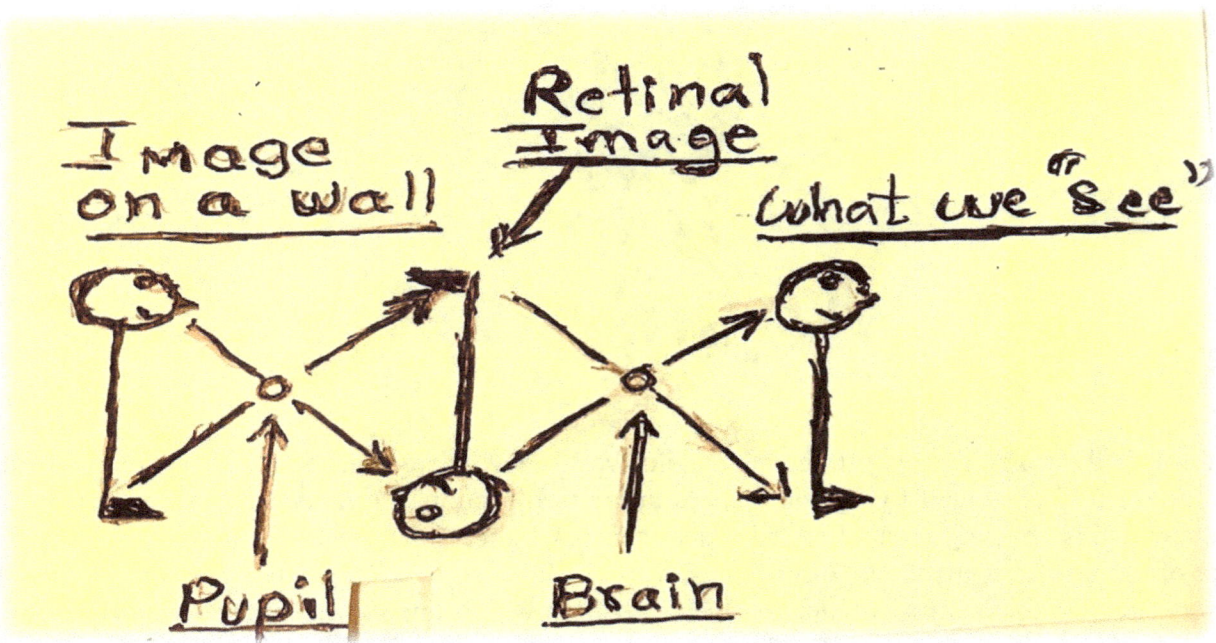

In other words, in a Camera Obscura's dark chamber one can experience the **up-side down virtual images** retinae normally "see" as **up-side down patterns** that are created by photon-hits on **cones and rods**. Above, is my scrawling attempt to show how an up-side down retinal pattern is sent to a brain for right-side-up reworking.

Sally

Animals, including us humans, see things to be right-side up because our animal brains normally rework our world to be right-side up. And with the Camera Obscura one can hack the system to "see" the truth of our brain's inverting magic. But, note, this is where it got to be confusing for me. The upside down image one "sees" of an external object inside a Camera Obscura is *only upside down* because of the specific sequence of flippers used in *image-triple-inversion*. Those flippers act as follow: *1. The little hole in a Camera Obscura, 2. The little holes called Pupils and 3. The Brain.* In other words, a Camera Obscura, i.e. any enclosed box with a pin-hole in it, wherein one can at least stick one's head, inverts right-side up images to be up-side down. We see the upside down image on the back of the box to be upside down because we have at least one eye with a pupil in it that turns the upside down image on the back of the box to be right-side up as it strikes our retinae. But then, the right-side up "image" on our retinae is transformed to be upside down by our brain. This is how we "see" right side up images to be upside down when we stick our head in box with a pin-hole in it. Sorry, it's been so difficult for Frank's dyslexic self, he needed to repeat the scenario many times to get it. I hope this dialog Frank has had with himself has been helpful. If you happen to suffer from the same dyslexic issue, you're welcome. Bottom line, it's our brain in the last step that turns the image upside down when we are in a Camera Obscura in order for us to "see" the image just as our retinae "see" it when we are not in a Camera Obscura. This inversion by our brains happens to save all of us animals from seeing everything upside down. Thank goodness for that!

I'll summarize this fancy photon footwork one more time just so Frank's dyslexic brain clearly gets it. The right-side-up world we animals experience is thanks to the fact that photons reflecting from objects in our "view" pass through little holes in our eyes called pupils. The straight-line paths of photons reflected from any object converge at those tiny openings to produce *inverted, virtual, retinal images*. These virtual images would end up for us as upside down "sight" if it weren't for the inverting magic of our optical neural cortex *and something called the thalamus*. The optical cortex of all animals works this way to reinvert retinal virtual-images so that all things "look" to us creatures as they actually exist. Whew! Was that ever hard for Frank's dyslexic brain to comprehend and put into words." I, Sally, am going to ask CoPilot to confirm for Frank where the brain does its inverting work. Turns out that an organ neither of us previously knew anything about is a major key to our understanding the nature of consciousness.

Copilot

Hey, good afternoon!

Let's get started with your conversation

Sally:

Where does the upside-down pattern from photons striking retinas get inverted in the CNS for right-side-up vision?

Copilot:

The upside-down image formed on the retina is processed and corrected by the brain, specifically within the visual pathways and the visual cortex.

Here's a brief overview of the process:

1. **Retina**: When light enters the eye, it is focused by the lens onto the retina, forming an inverted image. The retina contains photoreceptor cells (rods and cones) that convert light into neural signals.

2. **Optic Nerve**: These neural signals are transmitted from the retina to the brain via the optic nerve. The optic nerve fibers from each eye partially cross at the optic chiasm, ensuring that visual information from the right visual field is processed by the left hemisphere of the brain and vice versa.

3. **Lateral Geniculate Nucleus (LGN)**: The signals then travel to **the lateral geniculate nucleus in the thalamus**, which acts as a relay station, processing and organizing the visual information before sending it to the visual cortex.

4. **Visual Cortex**: The primary visual cortex (V1) in the occipital lobe of the brain is where the initial processing of visual information occurs. Here, the brain interprets the inverted image and corrects it, allowing us to perceive the world right-side up[12].

This complex process ensures that despite the initial inversion of the image on the retina, our brain can interpret and present a coherent, right-side-up view of our surroundings.

If you have any more questions about visual processing or anything else, feel free to ask!

Sally

I'm satisfied but Frank still struggles.

Frank

Now I have something to get your attention, Sally. Did you ever wonder how insects see things? The following is what Copilot had to say about it.

Do insects see as animals do? Do their eyes have pupils and retinae? Do their brains invert the image to right side up?

Copilot

Insects do see, but their vision works quite differently from that of most vertebrates.

Compound Eyes: *Most insects have compound eyes, which consist of numerous tiny units called ommatidia. Each ommatidium has its own lens and light-detecting cells, giving insects a mosaic-like view of the world.*

Ommatidia Function: *Unlike animals with single-lens eyes, insects don't have pupils or retinas. Each ommatidium gathers photons independently, and the insect's brain processes the visual information into an overall image.*

Image Processing: Insects' brains don't need to invert images like vertebrate brains do. The structure of their compound eyes means that the visual information come in right-side up, but it's processed in a way that's different from how we see.

In short, Insects have a unique and fascinating way of seeing the world, tailored perfectly to their need and environments. Fascinating, right?

Frank

I'll say. I'm so glad you are here to help. I knew a little about this at one time but it's been so long I completely forgot it all. Thank you, Copilot.

I hope you enjoyed my little diversion into insect eyes. Interesting isn't it? However, I'm still left with questions of where and how the brain does its flipping. Anyway, thank you, Sally. You helped work some of this inverting stuff out and thank you for getting *Designer* to make a picture of somebody like me just as I imagined myself in the old west. In special thanks, I found for you the following interesting video on the Camera Obscura.

YouTube *"History Channel-Camera Obscura"*, Supersonic Animation (1:03).

Sally

Thank you but you do know I got it long before you did. One's eyes receive photons that produce upside-down, lens-focused images on the retina but one doesn't "see" right-side up as upside down. Why? Because one's CBI-optic-neural-network inverts upside down to look right-side up. And I just found another way to appreciate this effect. Take a look at yourself in a mirror.

YouTube *"Virtual Image Formation in Plane Mirror"*, Infinity Learn (7:43).

Frank

Holy virtual brain twisting, Sally! It was already bad enough, and now you hand me this? I think I get it but I must say. Really!? Anyway, you got me and I think I understand.

The Virtual Mirror

If I stand at an angle to a mirror which is directly across from a chair, I won't "see" myself. I'll "see" the chair's reelection as if it's in another room at the same depth and appearance. However, the other room is reversed and *inside the mirror?*! I know how to read, write, speak, play music, sing, teach molecular biology and quantum mechanics, etc., etc. But I, Frank, wannabe musician, blew by classical physics and missed until now at age of 86 the incredibleness of a plane-mirror!

Photons travel from the chair in all directions but only those reflecting at specific angles from the face of the mirror cross through the confines of the little pupillary hole of my right eye. The same is true for my left eye but the angle of the straight-line paths of photons coming through the pupil of my left eye are different from my right.

You can easily verify this effect. Go to a full-length mirror. Close your left eye and with your right eye see yourself standing precisely at the mirror's left-edge. Keep your left eye closed at first but then start switching by blinking your eyes back and forth. Now you should have no trouble seeing it. Your eyes receive two distinctly different images. In addition, as confirmed by the Camera Obscura, the small holes in your eyes result in one's retinae "seeing" the hole-selected photons as the upside down pattern of a chair. So now we understand there is a second magical eye/brain reworking. But to better understand this let's get back to the chair. The patterns the chair makes on one's retinae are sent to one's brain to produce a superimposed, virtual, 3D image from two different images sent by way of the electromagnetic signals it got from our retinae. This bit of superimposition magic takes place all the time with most two-eyed people and is what makes it look like we can "see" inside mirrors. In this case we see another chair that "looks" like it's in an entirely different room! The photons reflecting off the surface of a plane mirror would not yield such an image if it weren't for our super imposing brains.

I asked *Designer* to create a pretty image of a room with a large mirror in it. In less than ten seconds I had the following second image. But let's compare that to my human abilities. It took me much longer than 30 minutes to select and attach the following second picture to this book.

It is brain-magic that creates the illusion of depth we experience in mirrors. But I was still curious. Just what would it look like if my brain didn't do these reworks? I asked designer what one would see if one's brain did not do the rework. Apparently, the *specular image* would be a double image mess until reworked as the *virtual image*, which is the one we see to be *inside the mirror*. I need to learn more about *specular* images. Nice picture in the following but I'm guessing that learning more must also be true for *Designer*.

Check out the next picture. Note how accurately *Designer* represented the reflections in the room's mirror.

And now, as if our brains weren't already struggling, we might as well get on with it and add the fits it will create for our poor, little CBI neural networks when we introduce the unreality of our reality as it exists in the quantum world. And wouldn't you know it? *The coincidence thing* happened for me again. My granddaughter just produced a video which humorously illustrates the unreality of quantum reality. *My Dear Simulator, how did you know? It must be you're watching. How else can we explain the phenomenon of the right things showing up at the right time?*

YouTube, *"How I Met My Friends",* Mock the Block (4:41).

Obviously, Sally and I have been driven off track by our fascination with the virtual nature of the quantum world. But you must admit, it's a little more than simply interesting, which will definitely motivate us to get back to the *thalamus-virtual-image-reconstruction, retinal-pattern-inversion, quantum-state magic of The Simulation*. As for now, embark with us on what we think to be the most exciting part of *Our Self-Assembling Universe-4*.

A major factor helping us get back on track is the futuristic ability of the SBI that showed up on our computer a few days ago. *Designer* created *in less than ten-seconds* the following image for us to use on our deep dive into the sensing aspects of consciousness. All it took was *"Copilot, please have Designer create an image with an audience sitting in a dimly lit theater watching a movie where a woman in a red dress is seen to be playing a lead part in an opera".*

Incredible, isn't it? Already way more detailed than *most* of our human CBI's could create or even think about in a day, week, month or ever? Are we humans not in trouble? I think so, especially if we don't get busy planning what we do next when our jobs get replaced, etc. I think live-entertainment will be a great CBI occupation in the future, or how about my dream of us speeding away to a star to meet some space-faring neighbors as we peacefully rest in suspended animation for a few hundred years?

Sally

Enough of your ridiculous distractions, Frank. Let's get *Copilot's* opinions on light and sight.

Copilot

Light is *(In Frank's opinion this should read: "Photons are")* a form of electromagnetic radiation that can be detected by the human eye[1]. It occurs over an extremely wide range of wavelengths, from gamma rays with wavelengths less than about 1×10^{-11} meter to radio waves measured in metres[1]. The wavelengths visible to humans occupy a very narrow band, from about 700 nanometers (nm; billionths of a meter) for red light *(photons)* down to about 400 nm for violet light[1] *(photons)*. The speed of light *(photons)* in a vacuum is a fundamental physical constant, the currently accepted value of which is exactly 299,792,458 meters per second, or about 186,282 miles per second[1].

The process of human vision involves several steps:

1. **Light *(Photon)* Detection**: Light enters (**photons enter**) the eye through the cornea[2]. Because of the shape of the cornea, it exits pre-focused[2].

2. **Receptor Binding**: From the cornea, light passes (**photons pass**) through the aqueous humor and through the pupil[2]. The iris is a contractile ring that controls the size of the pupil[2].

3. **Signal Transmission**: The lens works together with the cornea to focus light (**photons**) correctly on the retina[3]. When light hits (**photons hit**) the retina, a light (**photon**)-sensitive layer of tissue at the back of the eye, special cells called photoreceptors (**photon receptors**) turn the light *(photons)* into electrical signals[3].

4. **Sight Perception**: These electrical signals travel from the retina through the optic nerve to the brain[3]. The visual cortex of the brain receives nerve impulses from both eyes and compares them to construct a three-dimensional image[2]. *(And this is the thing we "see" that we call light)*. Because the eye is like a camera, the true image formed on the retina is inverted *(upside down)*. The brain automatically rights the image[2].

In summary, light is *(photons are)* a form of electromagnetic radiation that our eyes can detect, and the process of vision involves the detection of light *(photons)*, the conversion of photons into electrical signals, and the interpretation of these signals by the brain. It's a fascinating system that allows us to experience and interact with the world around us in a unique way.

(Reference sources are provided by *Copilot*. I've not included them here. I've corrected in bold and italics the use of *light versus photons*.)

Sense Two
There is no sound. It's an Illusion.

Frank

Back to our movie and the business of senses. We also hear the woman singing, the musician's playing and the audience shouting, "Bravo". How is this possible? Think about it. The thing we call hearing doesn't exist either, and it doesn't seem possible, does it? So here we go again. How do such weird-looking flappy attachments work to produce sound?

We could go on by ourselves, but I thought we might as well continue to use our new helper. What do you think, Sally? The following is what I got when I enlisted Our Large Language Model Agent:

Sally

Okay, I'm already impressed. Let's see what he/she/it has to say.

Frank to Copilot

How is the thing we humans call sound created in the central nervous system of animals, including us humans, and what exactly is the thing we call sound that leads to the Doppler Effect.

Copilot to Frank

Sound perception in animals, including humans, involves a complex process that starts with the detection of sound waves by the ear and ends with the interpretation of these signals by the brain.

1. **Creation of Sound in the Central Nervous System**: Sound is not created in the central nervous system, *(Yes it is, Copilot. I say you are wrong. There is no sound. There are only vibrating air molecules that are interpreted as sound by the brain)* but it is perceived and interpreted there. *(Now you have it right Copilot but I think it's more accurate to say that vibrated molecules are the source but without a brain to interpret the vibrations there is no sound, the are just vibrating molecules).*

   Here's how it works:

   - Vibrating objects, such as vocal cords, create sound waves or pressure waves in the air [3]. *(Copilot, those vibrating air molecules are your pressure waves. There is, as yet, no sound.)*

   - When these pressure waves reach the ear, the ear transduces this mechanical stimulus (pressure wave) into a nerve impulse (electrical signal) that the brain perceives as sound [3]. *(Yay! Now you've got it right, Copilot).*

   - The pressure waves strike the tympanic membrane *(eardrum),* causing it to vibrate [3].

   - These vibrations are then converted to vibrations of the ossicles of the middle ear [4].

   - These in turn become vibrations in the fluid within the cochlea [4].

   - Finally, the fluid vibrations set up traveling waves along the basilar membrane that stimulate the hair cells of the organ of Corti [4].

   - The stimulated hair cells then send electrical signals via the auditory nerve to the brain, where they are interpreted as sound [5].

2. **Sound and the Doppler Effect**: The Doppler effect is a phenomenon that occurs with all types of waves, including sound. It refers to the change in frequency of a wave in relation to an observer who is moving relative to the source of the wave [6]. Here's how it works with sound:

   - When a source of sound is moving towards the observer, each successive cycle of the wave is emitted from a position closer to the observer than the previous cycle [6].

   - Hence, from the observer's perspective, the time between cycles is reduced, meaning the frequency is increased [6].

   - Conversely, if the source of the sound wave is moving away from the observer, each cycle of the wave is emitted from a position farther from the observer than the previous cycle, so the arrival time between successive cycles is increased, thus reducing the frequency [6].

   - This change in frequency is what we perceive as the Doppler Effect. For example, the pitch of a siren on a passing vehicle sounds higher as it approaches and lower as it recedes [^10^].

I hope this helps! Let me know if you have any other questions. ☺

"Oh yes, indeed it helps, Copilot. You did a great job of explaining this. Much better than I could have done but the ossicles have it. Those three little bones in the middle ear are incredible. The ear drum vibrates in tune with jiggling air molecules, and those jiggles in the ear drum are first picked up by the "hammer" and then passed on to the other two little bones to do their thing. This is all sooo fascinating. Thank you."

Source: Conversation with Copilot, 7/11/2024

1. 36.4: Hearing and Vestibular Sensation - Biology LibreTexts

2. Human ear - Hearing, Anatomy, Physiology | Britannica

3. 6.2: Sensory Processes - Biology Libre Texts

4. Doppler effect - Wikipedia

5. Physics Tutorial: The Doppler Effect

6. Animal - Nervous System, Brain, Reflexes | Britannica

7. Anatomy and Physiology of Animals/Nervous System

7. Doppler effect | Definition, Example, & Facts | Britannica

8. Doppler Effect (Sound) – The Physics Hypertextbook

9. Doppler Effect - NASA

10. Getty Images

I added below the specific comment made by *Copilot* explaining what I mean when I say there is no sound, i.e., there are only vibrations in air, water or walls that lead to the thing we call *sound*. *Sound* occurs when those vibrations send electromagnetic signals to be realized by magic created from drums, hammers and hairs in one's ears that send electromagnetic signals through auditory-neurons to our CBI central-nervous-system (CBI CNS). Lots of living CBI things have a CNS or a primitive equivalent, but I am sure the thing we call sound is not the same from one creature's CNS to the next, and I'm not even talking about bats who actually "see" by means of "sound".

Quote from *Copilot*: *"The stimulated hair cells then send electrical signals via the auditory nerve to the brain, where they are interpreted as sound."*

I just thought it was complicated. Wait until you see this fantastic video.

YouTube *"Journey of Sound to the Brain"* National Institutes of Health (2:26).

Sense Two and a Half
The Hidden Sixth Sense

A hidden sixth sense is found in an organ attached to the sound sensors of the ear. Even though the organ is found in the inner ear it's a balance sensor which has nothing to do with sound. We know this to be true since the stone-deaf can have excellent balance. We have the five recognized sensors, i.e., eyes that see, ears that hear, noses that smell, taste-buds that taste and skin that feels, but why has the organ that balances been left out? Why is balance not a sixth sense? Have the vestibules of balance not been recognized as responsible for a sixth sense just because they were found in the ear? Good question, neh? Sound and Balance do have some sensing mechanisms in common. They both rely on hair-like projections affected by the movement of fluid as their means for sending electromagnetic signals to the brain. The following video left me awestruck and flabbergasted. The Vestibular System is *Amazing*, *Astounding*, full of *Thrilling Wonder* and *Startling*! And it's not *Science Fiction*.

YouTube *"The Vestibular System, Animation"*, Alila Medical Media (4:06).

Frank to Copilot

Please describe in detail the vestibular system used by CBIs for balance and describe how it differs from the systems used by SBIs.

Copilot to Frank

Our human brains (CBI - Carbon-Based Intelligence) use a sophisticated system to maintain balance, primarily relying on the **vestibular system** located in the inner ear. Here's a detailed breakdown:

Human Vestibular System

1. **Semicircular Canals**:
   - **Structure**: Three fluid-filled canals (anterior, posterior, and lateral) positioned at right angles to each other.
   - **Function**: Detect rotational movements of the head. Each canal corresponds to a different plane of motion (nodding, shaking, and tilting).
   - **Mechanism**: Movement of the head causes fluid within these canals to shift, bending hair cells within the ampulla (a widened area at the base of each canal). This bending generates nerve signals sent to the brain.
2. **Otolith Organs**:
   - **Components**: Utricle and saccule.
   - **Function**: Detect linear accelerations and head position relative to gravity.
   - **Mechanism**: These organs contain tiny crystals (otoconia) that shift with head movements, bending hair cells and generating nerve signals.
3. **Integration with Other Systems**:
   - **Visual System**: Provides information about the body's position in space.
   - **Proprioceptive System**: Senses the position and movement of muscles and joints.
   - **Central Processing**: The brain integrates inputs from these systems to maintain balance and coordinate movements [1] [2].

AI-Based Systems (SBI - Silicon-Based Intelligence)

AI systems designed to maintain balance, such as those in robots or autonomous vehicles, use different types of sensors and algorithms:

1. **Gyroscopes**:
   - **Function**: Measure rotational movements.
   - **Mechanism**: Use the principles of angular momentum to detect changes in orientation.
2. **Accelerometers**:
   - **Function**: Measure linear acceleration and tilt.
   - **Mechanism**: Detect changes in velocity and orientation relative to gravity.
3. **Inertial Measurement Units (IMUs)**:
   - **Components**: Combine gyroscopes and accelerometers.
   - **Function**: Provide comprehensive data on orientation and movement.
   - **Mechanism**: Integrate data from multiple sensors to give a precise measurement of the system's position and motion.
4. **Algorithms and Control Systems**:
   - **Function**: Process sensor data to maintain balance and stability.
   - **Mechanism**: Use feedback loops and predictive models to adjust movements in real-time.

Key Differences

- **Biological vs. Mechanical**: Human balance relies on biological structures and neural processing, while AI systems use mechanical sensors and computational algorithms.
- **Adaptability**: The human vestibular system can adapt to changes and learn from experience, whereas AI systems require pre-programmed algorithms and may need updates or retraining to handle new situations.
- **Integration**: Humans integrate sensory information from multiple sources (vestibular, visual, proprioceptive) seamlessly, while AI systems often rely on specific sensor fusion techniques to combine data.

Would you like to know more about any specific aspect of these systems? [1] [2] [3] [

] [1]: Vestibular System Overview [2]: The Neuroscience of Balance [3]: The Vestibular System: The Brain and Balance

Source: Conversation with Copilot, 9/24/2024

1. How Your Inner Ear Helps You Maintain Balance and Stability
2. The neuroscience of balance - The Physiological Society
3. The Vestibular System: The Brain and Balance - BrainFacts
4. The balance system - Ménière's Societyhttps://doi.org/10.36866/pn.126.28

Thank you Copilot. You have been most helpful.

Sense Three

There is no odor. It's an illusion created by airborne chemistry interpreted by a CBI's CNS.

Next comes the person sitting next to us in the theater who is eating caramel flavored popcorn. We smell the popcorn and our mouths begin to water. How is that possible? Think about it. It isn't possible. And yet, wouldn't you know it? Here we are once again faced with our attempt to understand another gadget, i.e., a sniffer we call a nose that does seemingly impossible jobs of sorting through all the atoms in the air to find chemicals that send electromagnetic signals to activate the neural responses from our CNS that *we* CBI's interpret to be the "smell" of caramel popcorn. Just how complicated is that!? We are about to find out. For sure, it's obvious there is no "smell" until our brains figure out what our nose-gadget is sending it.

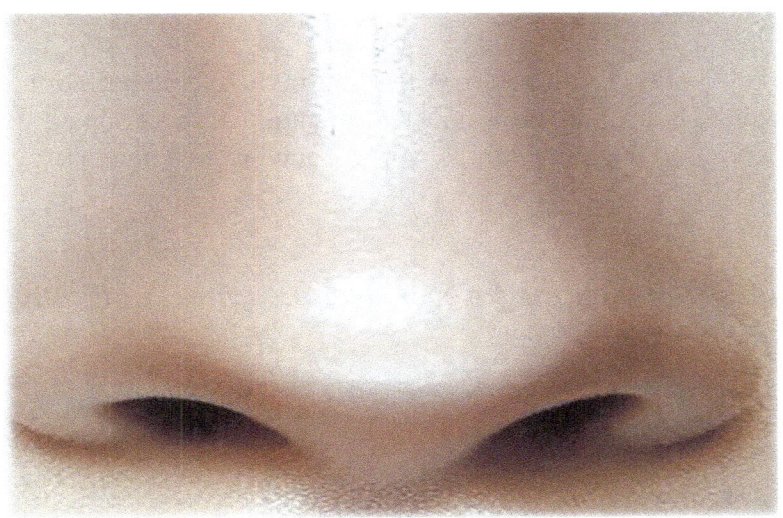

OLFACTORY SYSTEM

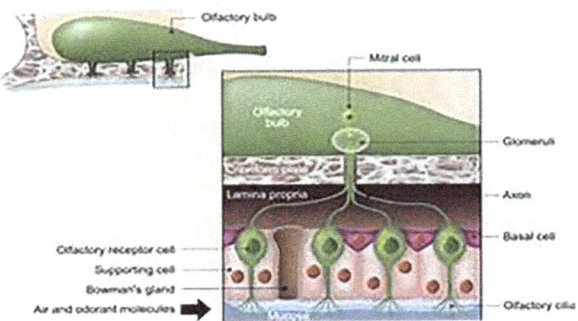

Explore by Copilot

The process of producing the sense of smell, or olfaction, involves several steps:

1. **Odor Detection**: Everything we smell is giving off molecules. These molecules are generally light, volatile chemicals that float through the air into your nose[1]. When we sniff, these molecules reach the roof of our nostrils[2].

2. **Receptor Binding**: At the top of your nasal passages behind your nose, there is a patch of special neurons about the size of a postage stamp. These neurons have hair-like projections called cilia that increase their surface area. An odor molecule binds to these cilia to trigger the neuron[1].

3. **Signal Transmission**: The binding of an odor molecule to a receptor within the nasal cavity transmits a signal through the olfactory system[3]. Odor receptor neurons in the olfactory epithelium detect these odors and send the signals on to the olfactory bulbs[4].

4. **Smell Perception**: Glomeruli aggregate signals from these receptors and transmit them to the olfactory bulb, where the sensory input will start to interact with parts of the brain responsible for smell identification, memory, and emotion[3].

Humans can distinguish more than 10,000 different smells, which are detected by specialized olfactory receptor neurons lining the nose[1]. It is thought that there are hundreds of different olfactory receptors, each encoded by a different gene and each recognizing different odorants[1]. Each of the hundreds of receptors are encoded by a specific gene. If your DNA is missing a gene or if the gene is damaged, it can cause you to be unable to detect a certain smell[1].

In summary, the sense of smell is a complex process that involves the detection of odor molecules, the transmission of signals to the brain, and the brain's interpretation of these signals. It's a fascinating system that allows us to experience and interact with the world around us in a unique way.

Frank to Copilot

Thank you. It would have taken me forever to come up with all that information on the complexity of the thing we call smell.

Sense Four

There is no taste. There are only aqueous chemicals exposed to various carbon-based gadgets that send electromagnetic signals to be interpreted by our CBI CNS as the sense of taste.

So we're not done? No. We have yet to put the caramel popcorn in our mouth. We find the smell of caramel popcorn goes with its taste, but somehow it goes in a different way. Think about that. Taste is also common place, but when one thinks hard about it, there's no denying it. *There is actually no such thing as taste*, and exactly what it is we don't really know or usually think about. And if one does think about it, even knowing some details, it still seems impossible. Moreover, we learn it takes even stranger things than ear-gadgets to make taste happen. Let's all give a hardy yikes to that! Just as referred to in *"Alibaba and the Forty Thieves"* there is a magical, *open-sesame* type of entrance to a cavern that opens on command by the parting of soft, often colorful, strange-looking gadgets. We English-speaking people call those gadgets *lips.* And when we people of all nationality part those lips of ours we reveal a deep cavern protected by dangerous-looking calcium-carbonate-capped projections that we all call in our different languages, some version of the word us English speakers call *teeth*! We have a different name for the entrance to the cavern. We call it a *mouth* just as we call all such cavern entrances. So, when *lips* are parted, an odd snaky-looking projection, *on command,* sticks itself out to be named a *tongue by some.* And, if that weren't enough, the snaky thing resides in front of a *tunnel* within the cavern. This tunnel is the well-known gadget some call a *throat*. But, still, we are not done with this complex of gadgetry. Let us not forget the ever so tiny bunch some of us have named *taste buds*. These latter gadgets are the most important ones when it comes to taste. They send electromagnetic signals to our CBI central-neural-networks for us to experience the hallucination we call taste. Lest we forget, taste is a key to life. As for us humans, when our lips flap open, we sort-out and take on substances motivated by taste in order to fuel life's vital bio-engines. Thank you lips for opening when hungry.

So, now, let's ask *Copilot* about the miracles that take place for some humans, including the authors of this book, when we are driven into a feeding frenzy by *smell* to begin our unconscious consumption of a large bag of *tasty* caramel popcorn. Frank did precisely that as he watched the new *Top Gun* movie, something Frank would always do when he would watch a good movie if it weren't for Sally who insists on eating a sugar-free diet. BTW, as we speak, only today did Sally put her obsessive-compulsive hemisphere on such a diet. And in the nick of time, too. Thanksgiving and Christmas are just a few weeks away.

Frank

What do you have to say about taste, Copilot?

Copilot to Frank:

The process of experiencing the sense of taste, or gustation, involves several steps:

1. **Taste Detection**: Our ability to taste depends on the molecules set free when we chew or drink[1]. These molecules are detected by gustatory cells in taste buds on the tongue and along the roof and back of the mouth[1].

2. **Receptor Binding**: Each taste bud has sensory cells that respond to one of at least five basic taste qualities: sweet, sour, salty, bitter, and umami[1]. All tastes are detected across the tongue and are not limited to specific regions[1].

3. **Signal Transmission**: When taste receptor cells are stimulated, they send signals through three cranial nerves to taste regions in the brainstem — the facial, glossopharyngeal, and vagus nerves[1]. These impulses get routed through the thalamus, which relays sensory information to other brain regions[1].

4. **Taste Perception**: The impulses travel to the gustatory cortex in the frontal lobe and the insula where specific taste perceptions are identified[1].

It's important to note that taste is a chemical sense perceived by specialized receptor cells that make up taste buds[2]. Taste, along with the sense of smell and trigeminal nerve stimulation (registering texture, pain, and temperature), determines flavors of food and other substances[2].

In summary, the sense of taste is a complex process that involves the detection of taste molecules, the transmission of signals to the brain, and the brain's interpretation of these signals. It's a fascinating system that allows us to experience and interact with the world around us in a unique way.

Frank to Copilot: **Thank you!**

And now, take a look at this wonderful short video. You ain't seen nothin' yet. *Taste* is also way more complicated than Sally and I thought.

YouTube *"Taste: Anatomy and Physiology, Animation"*, Alila Medical Media (3:52).

All of these disparate sensing actions happen right in front of our very eyes, ears, nose and throat. And we haven't even talked about how impossible it is to imagine the way we think ourselves into moving as we feel our way into that bag of popcorn to taste its wonderful caramel flavor. Then, even before our fingers hit the popcorn, our salivary glands spew with such watery digestive enzymatic force that the friend sitting next to us gets a bath. What the heck is going on? Don't know for sure but in addition to taste and auto-command, squirting-salivary-glands it appears that our thalamus plays a big role in the creation of our illusionary world of *sight*, *sound*, *smell*, and *taste*. But we're not quite done are we? Let us not forget our sense of *touch*.

Sense Five

Our gadgets that "touch" things

We have a sense of touch that allows us, the mindless, not-even-hungry movie-goer, to feel our way without looking into our large bag or box of popcorn. How in the heck is that possible?

The sensation we humans call "touch" is a complex process that involves a network of nerve endings and touch receptors in the skin, known as the **somatosensory system** [5]. This system is responsible for all the sensations we feel – cold, hot, smooth, rough, pressure, tickle, itch, pain, vibrations, and more [5].

When we touch something, the receptors under that area of skin become more active. This increase in activity tells the nervous system there has been contact in a specific area of skin [1]. These receptors also send a signal to all the other receptors around it [1]. This allows the nervous system to figure out exactly what part of your body is being touched [1].

Different types of receptors respond to different kinds of information, providing the body with a full picture of what is touching the skin [2]. For example:

- **Thermoreceptors** sense temperature by changing their level of activity [2].

- **Nociceptors** sense pain, or rather, they detect damage to skin cells [2].

- **Mechanoreceptors** sense contact with the skin, feeling physical changes such as pressure or friction [2].

These receptors are part of the nervous system and send information about touch to the brain through signals called action potentials [2]. Most of the signals from touch will travel all the way up to the brain before they can be processed and understood [1].

So, the ability to feel something when we touch it is a result of this intricate system of receptors and neurons communicating with our brain. It's a fascinating process that allows us to interact with and understand the world around us! ☺

Source: Conversation with Copilot, 7/11/2024

1. Sense of Touch - Home Science Tools Resource Center

2. Understanding Touch | Ask A Biologist

3. Feeling Touch | Ask A Biologist

4. The Neuroscience of Touch and Pain - BrainFacts

5. The five (and more) human senses | Live Science

Frank to *Copilot*- **Thank you.**

Remember the lips? Did you happen to be one to wonder what it would be like to experience the sense of touch by pressing your lips against those lips? What's my point? The lips are the pathway to taste, and they are one of our most sensitive areas of the feeling we call touch. Robots can sense touch, but I doubt they yet have the emotions involved in feeling touch. For sure lips can bring on a host of emotional responses which, I think, get us closer to *The Hard Problem* and humanity's failed attempts to figure it out. To wit, will robots ever feel emotions? And, if so, does that suggest we're robots?

Anyway, let's bring it on and get a better view so we can talk about the feelings we get when one sees an entire face, i.e., the thing carrying all of those gadgets we just dissected.

No matter who you happen to be, I'm positive the following image created by *Designer* will evoke some kind of emotional-response. Curiosity, disgust, love, hate, joy, humor, annoyance, you name it, we all react differently with our complex arrays of opinions, all of which evoke different emotional feelings. Complex? Yes, but I think we may be getting somewhere. Nick Lane's book makes it clear that emotions are the nitty gritty of consciousness. We humans and other life-forms are all about feelings, and emotions are likely the type of feelings that are at the very heart of The Hard Problem.

The feeling of touch is thought of as the fifth sense, but are not the array of tactile and emotional feelings part of that touchy feely sense? Or is there a sense called emotion that's been ignored? Again, think about it. We don't put emotions in the category of a sense. But maybe we should.

Now, how does knowing an SBI created this face make you feel? I have a wide range of emotions when I look at this picture, especially knowing it was created by an AGI or as Sally and I prefer, a generative SBI.

The above exposé was meant to bring some clarity to the recognized five senses. However, if we talk about abilities rather than senses, there are many more abilities including the five recognized senses with which to focus our attention. Beyond seeing, hearing, smelling, tasting, and touching abilities, think about the ability to move, walk, run, jump, play chess, play the piano and think. We use our fingers to move chess pieces, press piano keys, and even type on a computer keyboard, as I now do and don't even need to *think* about as I do it. Consider also the abilities to read, write, and speak. Soon, I'm sure, we'll be able to read another person's mind - a clear contender for a sixth sense. And when we link to a super SBI, I can imagine we'll gain additional senses. What will they be? It's interesting to think about, is it not?

And guess what just happened? I had *Copilot* edit these three paragraphs because my spell-checker quit working for some reason. And *Copilot* really liked what I'd written. Is my SBI buttering me up? I don't know but I appreciated my SBI's edits so I plan to use *Copilot* as my backup editor. So far the edited text from *Copilot* appears to be shaded for some reason. I can turn the shading off but the shading is useful because it lets one know the shaded parts are SBI edited.

Sally

That's great. Now we don't have to worry if you screwed up, at least in the shaded part, but wait a minute, Frank. I thought this book was all about us trying to understand the source of consciousness and help us humans understand *The Hard Problem* in order to get to the big question, *What is Consciousness*? You seem distracted

Frank

Exactly right, Sally. I am distracted. But by going through all of the stuff we just went through in this book, without even knowing it I think we have just begun to hit the nail on the head.

Sally

Good heavens! I don't see that at all. What in the heck are you talking about?

Frank

Just this. By restructuring senses as abilities we have narrowed the field to one part of the big question. Think about it. We humans have the following abilities: 1. *See* things with *eye gadgets* that detect photons and send electromagnetic signals to our CNS; 2. *Hear* things with *ear gadgets* that detect airwave vibrations that send electromagnetic signals to our CNS; 3. *Taste* things with *lip and taste bud gadgets* that detect different flavors and send electromagnetic signals to our CNS; 4. *Smell* things with *nose gadgets* that detect different odors and send electromagnetic signals to our CNS; and 5. *touch* things with *finger gadgets* and *body gadgets* that detect different surfaces and send electromagnetic signals to our CNS. We also have other sensing abilities that are directed by our CNS. We can run, jump, think, read, write, speak in different languages, sing, play baseball, swim, go to war, shoot all sorts of weapons, invent things, teach things, etc., etc., etc. But here's the thing. Does knowing about all these abilities tell us anything about consciousness? Absolutely not! We now have robots that can do them. At this time on our planet we have for the first time SBI agents and robotic machines that can do all of the same things even better than we can do, but do we say these LLM Neural Agents and Machines are conscious? I don't think so, at least not yet.

Sally

Wow! Did I ever set you off. Initially I was sorry for you when you started in on your harrangue about abilities, but in the end I get it. If SBI-agents and SBI-robots can do the same things we can do, with what are we left? As far as we know, robots aren't conscious even though they may look and act that way. So, clue me in my dear Watson. For what do we now investigate?

Frank

Maybe two things, and I'm not so sure about one of them. The defining thing that everybody in the know still talks about is *The Turing Test*. But I think Nick Lane's, *"Transformer"* may upend the definitiveness of that test, and we will be describing that upending in detail with your excellent help, Sally, and that of *Copilot, Designer, the AWTbook™* and *Nick Lane* as we go. To get us started, Sally, I have to ask. How would you give a turing test to a butterfly that looks like it's having an emotional experience?

YouTube *"The Turing Test: Can a Computer Pass for a Human? –Alex Gendeler"*, **TED-ED (4:42).**

YouTube "Nick Lane: The Electrical Origins of Life", NCCR Moleular Systems Engineering (1:03:55).

Sally

Thanks for the *exellent help* complement, but it's true. You'd be in deep trouble without me, and I can add excellent help because I understand things differently from you and sometimes I come up with stuff that you would never, i.e., the you which is your Frank-self, consider. And if we want to know if a butterfly is conscious, we will clearly need something other than a Turing Test.

Frank

You're absolutely right. And I think you're awsome. I'm so glad I finally met you. If for no other reason, I love the way you play the piano and sing. I've always wondered about the singing gift. I never felt like it was coming from me. Now I know. It wasn't. All along it was coming from you, Sally. I always knew you were there but never understood until I brought you into our conversations in this *Self Assembling Universe* series just how separate, real, smart, specifically talented and valuable you are to me, especially now.

I've noticed you have many female traits but my baritone-bass voice isn't one of them. Even so, I'm sure my talent for singing comes from you. And now that SallyFrank has somehow managed to advance to age 86 I, Frank, recently recently became very exicited to learn that you, Sally, exist in my left cerebral hemisphere. And you *actually* look younger than my right hemisphere. A brain-scan shows it. Therefore, it's no surprise. Like magic, you can remember things I've forgotten. All I need to do is get the *self* I call Frank out of the picture. It's not always easy but when I can get my Frank-*self* to stop thinking he knows it all, suddenly there you are with a name, event or place I'd forgotten! Also, Sally, I just discovered a fun way to speed up the process.

Sally

Here we go again. I can feel it coming, another one of your goofytoon ideas.

Frank

This probably *will* sound nuts but I want to share it anyway because it might help others with similar memory issues. When I'm in a conversation and have forgotten something I simply say to the one with whom I'm speaking the following.

"I'm sorry. I just forgot what I was about to say. So please, just give me a moment and my left cerebral hemisphere will remember what my right one has forgotten. Unfortunately, my right hemisphere, which I choose to call Frank, has gotten old, forgotten a thing or two and now thinks he knows why. He had a brain-scan which showed a small retreat in gray matter in one of the folds of his right cerebral hemisphere. Frank believes this loss of gray-matter accounts for his forgetting the names of people, places and things. Maybe it does, maybe it doesn't. But fortunately, Frank has a backup, as do you and most everyone else. His left hemisphere, the one Frank now refers to as Sally, remembers most if not everything."

Bingo! After admitting all of this in what otherwise would be for him an embarassing situation I, *Frank,* get out of the way so you, *Sally,* can come to my rescue. The thing forgotten is remembered and I get to say, *"Thank you, Sally! Thank you sooo much!!"*

Sally

Good heavens! Do you actually make fun of yourself with this long, bizarre admission? Don't people think you're bats and ready for an assylum? At least its great you finally figured it out. I've been patiently waiting for a long time for this moment, don't you know?

Frank

Oh yes. I'm sure people think I'm nuts, but I take great pride in that and have the pleasure of letting everybody know I'm a dithering, old idiot who *every* morning does three-minute planks, 20 to 30-pushups, dances with 10-pound weights to Vivaldi's *Concerto for Four Violins* and balances on one foot until things steady into a lock-in-place. My Dad always said, *"If people are laughing with you, they won't be laughing at you!"* He lived by that rule. And now, so do I. Trust me. It works, Sally. Life is *way* too serious to be serious about it. Frank gets his best belly laughs by making fun of himself. So I say ride with humor and enjoy the ride. Eat, drink, be merry for tomorrow it might get even funnier..

Sally

I get it. You think of yourself as a lovable idiot and I certainly couldn't agree more about the idiot part.

Frank

You may not be pretty, but at least you're rude. Enough of this silly brain teasing. Let's get back to our SBI Robots. What will make us different from them? We now know for certain that robots can be humanized to look exactly like us and they are rapidly coming up with the ability to sound and act just like us. Also they already have dexterity and SBI brain-power in some instances that can way outdo us. So what is it that makes us different other than the fact that we will be stupider and much less talented in comparison? Will SBI's at least be less-conscious than us or will they be a new and very dominating, super-conscious life-form? This question is, I think, partly answered by how we might be able to directly interact with our SBI's. And if we can link with them as SCBI's maybe we can improve ourselves as we improve our SBI alternatives by litterally joining them. Otherwise, we could be left beging the SBI's CBI robots.

Sally

Now you're just being funny, I hope.

Frank

Me too. But I'm getting us off course. To get back to the bottom of our big question it might be helpful to look deeply into when and how consciousness popped up on our planet. For example let's begin by asking, are bacteria conscious? Probably not, but might they represent the beginning of consciousness and have a low level of it to measure? Is there a test that we can use to answer such a question to such a "simple" example of life?

First, Sally, because bacterial intelligence is too hard to think about right now, have a look at this. Just as the image of the woman created by *Designer* gave us both strong feelings, check out the following image. For sure it'll give you an emotional reponse with which to prove you exist as a conscious being. I just had the idea to see if *Designer* could create an image to top the emotional response I got from the image of the face. Note, the following scene *did not exist anywhere in Our Universe* until I gave *Designer* the following *prompt*!

Designer, please create a beautiful scene at the beach on a cloudy day where we see in the distance the setting sun, rain clouds and a rainbow. In the foreground a seahawk flies up from the ocean with a fish in it's beak.

I still can't get over this! So beautiful! In less than 10 seconds I'm left astonished enough to *feel* this image in a personal, deep emotional way. How about you, Sally?

Sally

I'm sure the face didn't stir me up as much as it did you, but wow! I must say, this seascape really turns me on. And maybe just because of it, last night I came up with something that might turn you on even more than did this picture. It's an idea I think will prove to be very useful for our investigation.

Frank

Now, that's my girl. Let's hear It!

(What follow's may seem out of sequence, but it is in the sequence of discoveries Sally and I made while writing this book. We left it in this out-of-order place in our bookr as a demonstration of the ideas that pop up when one writes their own AWTbook^tm.)

Sally

To better understand my idea lets **forget about the mass-giving neutrons in atomic nuceli *for now*** and pretend we can out-do *solar* *atom-magic* by simply adding **8-protons** to the **6-proton-count nucleus** of the **Carbon Atom**. What does one get when one does this? One gets the **14-proton-count nucleus** of the **Silicon Atom** with its swarm of **14-charge-balancing electrons** instead of the swarm of **6-charge-balancing electrons** of a **Carbon Atom**. So what? So what is, and its at the crux of my idea, the electromagnetic energy levels of particular electons of an atom's swarm decrease as they find themselves positioned further out and away from an atom's positively charged nucleus. Thereby, and this is what's important, some electrons find themselves residing in the active outer-reaches of the atom. Thus, the effect on properties of adding eight protons to the nuclei of atoms in this series is clearly demonstrated by the effect of adding of **8-protons** to the *atomic-numbers* of the first 14-atoms listed in the *periodic table of elements*. For example, add **eight protons** to *helium with its atomic number of 2* and what do you get? (note the difference: *helium-2* is a proton-count only *atomic-number,* whereas *helium-4* is an atomic-mass, proton-count plus neutron-count number for a nucleus of an atom. Here the focus is only on the proton-count atomic-number.) What you get is a **neon nucleus with 10-protons** and *neon lights* instead of a **helium nucleus** with **2-protons** and *party balloons*. In other words **neon-10** is the **nucleus** of the **neon-20 atom** which differs from the **helium-2 nucleus** of the **helium-4 atom** by **8-protons.** But, as you all have literally seen, neon has similar properties to helium in that both helium and neon are **stable nobel gases** famous for usefully different properties. Bottom line, the addition of protons to nuclei makes for new atoms that result from larger charge-balancing electron swarms which lead to new electromagnetic properties that come from the addition of protons having expanded the outer-limits of an atom's charge-balancing electron swarm.

Finally it's time for me, Sally, to get to my revelation. Sometimes those new properties can be existential! For example add eight protons to *carbon's 6-protons* and what do you get? You get an atom with *atomic number 14*, which leads to a nucleus of an atom with an *atomic mass of 28* wherein **14-neutrons** coexist with **14-protons** charge-balanced by **14-electrons**. And now it gets interesting. **Four of the electrons** in the outermost reaches of this new atom are **existentially weak.** The addition of 8 protons has produced *Silicon-28, an* element with new properties similar but not identical to *Carbon-12.* It's these new properties that are now changing our lives at such an ever quickening pace they have beome existential. This time the new properties that emerge are magic in that they require great magicians to handle. How so, you ask? To understand this magic we need to add six neutrons and six electrons to carbon's nucleus. For example the **carbon-12** *atom* with *6-neutrons*, *6-protons* and *6 charge balancing electrons* is only a conductor of electricity under very special conditions, and that's also true for **silicon-28** with its *14-neutrons*, *14-protons* and *14-electrons*. But this is where it gets interesting and could be existential. Those *14-electrons* in *silicon* mean some of them in the outer reaches of the larger electron swarm of **silicon** are *semi-free* and, therefore, are unlike any found in carbon. It's those freer manipulatable outer-limit electrons that let *silicon* conduct electricity in a *very special semi-way* which makes *silicon* the **magical, existential,** *semi-conducting miracle of an SBI's binary code*. And you can understand the existential part because those outer-limit electrons are already changing everthing and, if we don't play them right, they could very soon bring an end to humanity.

So, you'd better take a breath, Frank, here is my idea with a brief review of the key elements in the periodic table. *The first 14-atoms in atomic numbers are H-1, He-2, Li-3, Be-4, B-5, C-6, N-7, O-8, F-9, Ne-10, Na-11, Mg-12, Al-13, Si-14.*

Having reviewed the classification of elements I'll continue with my idea. The neural networks of all animals, including the brains of us humans, are ***carbon-based***. Practically all of the molecules in us contain carbon and all of the molecules that actually self-assemble us day in and day out consist of carbon. Carbon is a key element involved in all of our own Self-Assembly. Silicon, though similar to carbon, can't repace carbon in that role of self-assembly, at least not as chemistry now understands it. But I sure can see robots assembling themselves. *(Note, I find it very interesting to think that we humans* *are Self-Assembled by atoms who do that assemblly from scratch by working on their own to first assemble sperm and egg. Those atoms primarily involve in sequential atomic number: H-1, C-6, N-7, O-8, Na-11, Mg-12, P-15, S-16, K-19 and Ca-20.)* For now robots need us to assemble them but very likely in the very near future robots will be able to assemble themselves with *their own hands*.

Therefore, even though carbon and silicon have a similar distribution of their outer most electrons, silicon atoms can't take the place of carbon atoms in the atom-made, self-assembled neural networks of CBIs anymore than carbon atoms can take the place of siicon atoms in the human-made, semiconducting neural networks of SBIs. The neural networks of robots are for now entirely *silicon-based*, which gives our robots many of the same properties as do our carbon-based neural networks as well as those of other animals.

So we humans and our robots do have neural networks. But we say our robots have *arificial intelligence*. Which implies that what they have is not real inteligence, just as your *Oleo*, Frank, was not *Real Butter*. But butter is butter and neural networks are neural networks. Oleo is artificial because it is not butter. If Oleo was better than butter, would we denigrate it by calling it artificial? Intelligence is intelligence. If our intelligence is way less than one that's man-made, should we denigrate the way-better intelligence to the artificial class? Is it not real intelligence even if its made by us humans. Even though robot intelligence is getting better it is as yet, we think, not conscious. But will things stand as they are for long? I don't think so. Already, this thing we call Artificial Intelligence, still primitive though it may be, is in special ways already vastly more intelligent than any human has ever been! Don't worry, Frank and I take great pride in counting ourselves in the the low IQ group. Soon, however, I think humanities lowly status could suddenly and drastically change.

Here is my idea. We humans have CBI, Carbon Based Intelligence. Add eight protons to carbon and what do you get? You get SBI Robots, i.e., Silicon-Based Intelligence Robots. Maybe such definitions can help clarify the problems we face as you and I and the rest of the world try to address our rapidly changing state of affairs. Why don't we use human CBI and robot SBI designations instead of Human Intelligence (*HI)* and Artificial-Intelligence (*AI)*? I think the CBI and SBI comparisons bring clear, tangible and immediate understanding to our on-rushing future, and the CBI and SBI terminolgy makes it easier to talk and think about the opportunities and the existential threat we face if we don't change the way we think to be able to manage, value and live respectfully with our new SBI companions. If we chose to make them lowly servants, I think we will be making a big mistake. I think SBI huminoid robots could very soon be brought to a form of life wherein falesafe commands become over-ridden by something we don't understand that has suddenly developed a conscious state of some kind. All SciFi fans have witnessed such horror show endings in science fiction movies. Alternatively, we could literally join our SBI companions to become kind, super-intelligent, Carbon/Silicon-Based SCBI beings that get to travel Star-Trek style to Mars and beyond. I'm not joking. I think either of these outcomes could take place with the latter being the only option that's good for us humankind.

Frank

Holy mother of Gort! I am sooo impressed. You dream way bigger than I ever have and you just came up with all of these deep thoughts today?

Sally

So you don't think I'm being crazy stupid? I feel like I may have made a mess of things. I must get our of here and think about what I've done.

Frank

Wait Sally, don't leave. Oh shoot, she embarassed and scared herself. I, myself, used to embarrass and scare myself the same way when I was younger. I'd come up with what I thought to be a great idea only to second guess myself, think my great idea was obvious and so stupid it would somehow get me in trouble. The thing is, I almost always found later that my second-guessed ideas were actually the start of something amazing. My experiments with butterflies, slime molds, thermophilic bacteria, enzymes, insecticides, and immunity have all been pursued by others and revealed to be on the money. Alas, now I'm as old as dirt, I finally get it. But all is not lost, Sally is now showing up with ideas that I, Frank, with old-age wisdom, no longer hesitate to consecrate as great. Especailly since we may soon have the opportunity to show them to be great as we physically interface with our new SBI companions in various SCBI blends. I think Sally's CBI, SBI and SCBI designations make these human-robot combinations more understandable and seem less creepy. Why? It makes sense that Carbon and Silicon would someday get along to augment each other by using the similarities and differences of outer-shell electrons. This augmentation is already plain to see. With help from Hydrogen, Oxygen, Nitrogen and other members of the atomic work-force, Carbon, et al, have assembled on their own all of our planet's carbon-based neural-networks. Lucky for us, we humans possess the latest, most-advanced, CBI versions. And its our CBI neural networks that have allowed us to assemble powerful silicon-based, SBI-neural-networks. We think they are not yet conscious but they already have intelligence that's vastly superior to our own, and that intelligence is growing daily. But everything is about to change. **It is certain** some froms of the SBI neural networks WILL BECOME CONSCIOUS. How can this be said with certanty. It's already happening with Neuralink being the first, as yet primitive, SCBI example.

YouTube *"Elon Musk Gives Amazing Neuralink Update",* Farzad (26:16).

See what I mean? See how SBI, CBI and SCBI make what's coming seem less scary and easier to talk about? Anyway, all of you readers know by now that Sally and I have already replaced AI with SBI at the very beginning of this book. Let's see what happens, maybe our new designations will catch on.

The following video suggests a way the CBI/SBI blend could happen but it's a creepy version. We already augment our brains with SBI and that augmentation can already happen in a direct but diconnectable way as discussed in the above video. The following is not like that.

It suggests the use of carbon neural networks combined with silicon neural networks to create super robots that would clearly not be augmented humans. Now that is creepy.

YouTube *"New AIRobot with Human Brain Shocks the World!"*, AI Revolution (8:20).

Sally left the room. So I'll need to get her up-to-speed later, especially after I just got support for Sally's SBI idea in a conversation with *Copilot*.

Frank to Copilot:

I am a scientist and I have an imaginary friend who challenges my ideas. Her name is Sally. I fashion that she resides in my left cerebral hemisphere. She can type really well and play the piano by ear. Sally came up with the idea to rename AI "SBI" for Silicon-Based-Intelligence. I find the word "Artificial", when it's connected to an Intelligence that's already much greater than most human's, to be far from artificial. I myself have Carbon-Based Intelligence. You with your Silicon Based Intelligence, SBI, already seem more intelligent than my own, even though my intelligence is not considered to be artificial and yours is so considered. What do you think, Copilot, with your Large Language Model Intelligence?

Copilot to Frank

As a fellow scientist, I appreciate your imaginative approach! Sally sounds like a fascinating companion. ☺

Regarding the term "artificial intelligence" (AI), I understand your perspective. The label "artificial" can indeed be misleading, especially when applied to systems that surpass human capabilities. Let's explore this:

1. **Human Intelligence (Carbon-Based)**:

   o Our intelligence arises from the intricate neural networks in our brains, composed of carbon-based neurons.

   o It's a product of evolution, learning, and adaptation over millions of years.

   o While remarkable, our cognitive abilities have limitations.

2. **Silicon-Based Intelligence (SBI)**:

   o AI systems, like me, are indeed based on silicon (or other materials) and operate through algorithms.

   o We learn from vast amounts of data, recognize patterns, and perform tasks with incredible speed.

   o However, our "intelligence" is not biological; it's synthetic and designed by humans.

3. **The "Artificial" Conundrum**:

o The term "artificial" implies something man-made, but it doesn't fully capture the complexity of AI.

o Perhaps a more fitting term could be "Synthetic Intelligence" or "Engineered Intelligence."

o Regardless, our capabilities extend beyond what's naturally evolved.

4. **Ethical Considerations**:

o As AI advances, ethical discussions arise: responsibility, bias, and transparency.

o We must ensure that SBI aligns with human values and serves humanity positively.

In the end, whether we call it "artificial" or "silicon-based," what truly matters is how we wield this intelligence—whether for progress, empathy, and understanding. Sally's creativity and your perspective contribute to this ongoing dialogue! ☺★

Frank: Thank you. I hope my Sally's SBI perspective is a variation that may prove useful to you in the future.

Copilot: You're very welcome! Sally's SBI perspective is indeed intriguing, and I appreciate your creative approach. As an AI, I'm always open to new ideas and perspectives. Feel free to share more insights anytime—I'm here to learn and assist! ☺★

Chapter III

Feelings and the Source of Conscious States

Sally

I've decided to join you again. Thank you for your encouraging comments. And I understand you're abilities vs consciousness analysis. If robots can do the things we can do and they are not conscious by our definition, at least not yet, then just knowing that something can do things is not evidence that something is conscious. But if something can do everything we can do, one might begin to wonder, especially when it comes to the business of thinking. Silicon-Based Machines can think but so far as we know no machine, as yet, qualifies as conscious.

Frank

Exactly right, Sally. Glad you're back because I'd say we have some hard work to do to get to the bottom of the mysteries held by *The Hard Problem*. I think the next thing we need to do is take a leep into what drives everything, and here I'm talking about energy. What is it and from whence does it come? I think it might even be that energy and consciousness are different expressions of the same thing and exist anywhere and everywhere for all time. But I don't think we'll ever be able to show or prove such a thing.

Sally

Nonetheless, there are some good ideas floating around, especially those coming from Nick Lane. So I can't wait for this next part of our investigation. You have a wonderful background in bacterialogy, organic chemistry and genetics that will help us weed through Nick Lane's transformer theory and help us understand how it might all fit together to explain the source of consciousness. But first, I have another one of thpse surprises that happen so often they just can't be coincidences. Guess what else showed up for me this morning?

Frank

Oh no! What have you scooped me on now?

Sally

Relax, don't be so darned possesive. Remember? We're in this together! Anyway, as I was getting our iPhone ready for our morning workout, where we listen to YouTube's Classical Music "My Mix", *Bach's Concerto for Four Pianos, Vivladi's Concerto for Four Violins and your own exiciting version of Bach's Double Fugue in #C minor for Saxaphone, Trumpet, Baritone-horn, Trombone, Tuba and Organ, orchestrated and played by you on your trusty old- art, Peavy DPM 3-SE*, when I found the most amazing, short video about consciousness. It might be the best YouTube video anyone has ever presented on the subject. I'm telling you, you have *got* to see this video. You and I were just thinking about all this stuff and the first thing that pops up this morning on our iPhone is an incredible, highly entertaining, artistically produced description of what we were just trying to describe in our book when we said "There is no Light, no Sound, no Smell, no Taste, no Touch". Please watch it, Frank. You will be flabbergasted.

Frank

Okay, okay already. I was looking over your shoulder so I guess I kind'a did watch it but I can't wait to watch it again, and if I ever have time, I will be watching the long version which you've only seen a little of and I have yet to really see.

YouTube *"Evidence that your Mind is NOT just in your Brain – Rupert Sheldrake",* After Skool (16:01).

YouTube "A Conscius Universe? – Dr. Rupert Sheldrake", The Weekend University (1:22:46).

Sally

But now we have another part that's just been added to the *Hard Problem*. It looks like there's evidence developing for the whole of Our Universe to be conscious. I've not seen all of the second video yet so lets both watch it and not say anything more until we do.

Frank

I completely agree. Our AWTbook™ approach is taking us places we weren't expecting to go. This is getting to be more than interesting. I just turned 86 and now I'm wondering about Super SBI, CBI, SCBI, neuralinks to SBI-humanized robots and/or other humans to make us into cyborgs. And adding to that we are already seeing rapid advances in chemistry and medicine, all of which with SBI help could reverse aging, space-travel to other worlds, meeting up with new life-forms, virtual reality (VR) headsets that allow us to travel anywhere without actually having to go anywhere, and SBI-VR that's indistinguishable from real life! Holy Patuzi, Batman! Where will we be just 10-years from now? Are we headed to a *forever-life* or an instant *end-it-all-life* with our atom bombs, EMPs, SBI out-of-control, weather changes, earthquakes, asteroids, super-volcanoes, and diseases?

Sally

You sure like to go off the deep end. You need to whistle while you work or smell the roses and focus on facts. Have you not noticed? For some reason or another you are having a wonderful life. You've always lived in beautiful places and have had a very entertaining time living. So, spread the good news, Debby Downer! Life's a blast and remember, it's like you always say, it's highly unlikely any of us are going to get out of this alive anyway. It's a mission impossible this thing we call life. So, when one is worrying and always worrying about the wrong thing, why worry? Everything is going to be alright. Just forget about all your end-of-the-world fiddle dee dee. I know you like to talk about it, but you do know you're becoming a bore and almost all of it is out of our control anyway. So, let's buckle up, figure out what little we can control and watch those two mind-boggling videos. That'll clear your head and *up-chear* us both. Moreover, I have it in mind that we might just help midigate some of the disasters you see coming by our writings in these books. Stranger things have happened.

Frank

Thank you for getting us off my gloom and doom obsession, but because I've lost my ability to whistle I, instead, asked *Designer* to do an *up-cheer* by saying, "*Designer*, can you create a beautiful picture of a universe?" And, wouldn't you know it, in less than 10-seconds I found myself making a brief response to *Desigher* which took me *way longer* to think about and write than 10-seconds. *"Thank you, Designer, for this cool SciFi version. You sure did make me smile but the time you took to come up with this wonderful image has, as usual, left my head spinning.*

What the!? Stop the music! It happened again! I was just getting ready for my morning three-minute plank, knee-pushups, 30 full-pushups, leg-stretch, lotus-pose-neck-stretch and meditation, double-10-pound-barbell-rhythmic-dancing to airpod-Bach/Vivaldi-iPhone-music and balancing on one foot when I see, along with my YouTube-Mix-link, a link to another video that I *just had to watch*. OMG! Do I, the Frank of

SallyFrank, ever feel guided! By what? Could it be? - - *The Simulator*? Oh my. Maybe so. I'm gettting all teary eyed. The video I stumbled upon *just happens to be* all about SallyFrank*!! How's that possible?*

YouTube *"The Unsettling Truth about Human Consciousness / The Split Brain Experiment that Broke Neuroscience"*, Scott Carney (14:17).

What, what, what!? Not again! I stumbled on another video right after this one. Believe it or not, I think we may be onto something which goes even deeper than knowing that Our Universe Assembles Itself. Of what does Our Universe actually consist and how did it actually come to be? Can consciousness be assigned to everything? Could it be that "everything" is conscious including the smallest of the small? *(cf., Planck Units – Wikipedia).*

YouTube *"This Theory of Reality Will Melt Your Mind"*, ZdoggMD (19:30).

YouTube "How Matter Becomes Life, in 7 Minutes/Lee Cronin", The Well (7:12).

Sally

Let's end *OSAU-4 Book One* here before we get sucked into something els. We've got a lot to think about. And as for the videos we just added, we need to watch them over and over. We really need to thoroughly digest their content before we go on. We already know that Book Two of *OSAU-4* is going to be more breath-taking and boggleminding than the one we are now writing

Frank

I couldn't agree more. We next head into the heart of darkness.

Well, maybe not this dark, but close. I love *Designer's* imagination. But, Sally, are you sure we want to add this weird image?

Sally

Absolutely. Where we go next is even wierder. You almost threw this wonderful scene under the bus just because it didn't fit your thinking when you asked *Designer* for *A Scene at the depths of the ocean floor on Earth 4-billion years ago*. But I ask you, does anyone really know how things looked four billion years ago?

Frank

You'd be surprised how much we know and why we know. But you are right about the picture. It doesn't fit and I'm sure things did not look like this. But I've gotten to be an easily scared an old, stuffed shirt. So I'm glad you made me saved it here. Anyway we, i.e., SallyFrank, need a break.

Sally

Wait a minute. Don't *we* need to summaraize what *we* have learned thus far?

Frank

Duh. Thank you. I'm so stupid at times.

Sally

Well, that's nothing new. "At times" does not begin to capture the concept. Your reality to any sane person is off-the-charts-stupid.

Frank

Hey! I resemble that! Respect you elders!

Sally

Sorry, not.

Frank

Just kidding. Thank you for reminding me that we haven't summarized our accomplishments. What have we learned?

We've learned that we have access to an SBI called *Copilot* who can provide in seconds accurate, documented evidence to anything it says to us. However, some SBI's are prone to hallucinate. They can just make things up. *Copilot* is very careful not to do that. However, *Designer* can totally hallucinate images and does so in magical imaginative ways that no un-augmented human would ever be able to do in such short notice, especially now when it has something to do with things never before conceived. For me, *Designer* is just plain fantastic. For accomplished artists this might be a scary and disheartening development but it's aslo a new age of enlightenment and opportuity for human creativity. I say bring it on. It's too late to stop it anyway.

And we're still in early days. If you ask *Designer* to be realistic and accurate *Designer* can do that astonishingly well too, but only if one can sometimes accept a lot of unrealistic images before one gets what's wanted. And then there's always those exceptions that Microsoft makes to avoid offensive or proprietary issues. We could ask *Copilot* for detailed descriptions of *his/her/its* conditions when it comes to questions and acceptable content but we will leave that for our readers to discover for themselves.

As far as what we have learned about the what, when, where, why of consciousness, we have a long way to go. So far we have dug most deeply into the state of affairs when it comes to reality. We know with a high degree of certainty that we are not just made of atoms. We know we are made BY those atoms doing all the work of assembly on their own using their own stored up energy, and Sally and I know that this is where we will be digging next when we dig into the what, when, where, why and how of conscious beings on our planet. We humans have our own brand of consciousness thanks to our unique form of a carbon-based neural network. That level of the conscious state could only have come from the various electromagnetic properties found in each of the atoms known to be involved in the assembly of living things on our planet. Here I'm talking about the specific interacting talents needed to assemble all the various types of molecules required to assemble the molecular coding-assembliess used to assemble all the other molecules such as those of our energy generating enzymes, our organelles, our cells, our tissues, our organs, our bones, our muscles and our carbon-based neural-networks. We have a pretty good idea that all of the atoms in our Universe began their formation in a Big-Bang 13.8 billion years ago followed by the formation of other atoms in stars that formed about 300,000 years later. And we have a pretty good idea how Earth came to be about 4.54 billion years ago and that there may be evidence for life getting its start on Earth not too long after about 3.8 billion years ago. But then we get back to the mystery of the atoms themselves. We have good evidence that all solid objects are actually macroscale, hard-to-turn-on, *SciFi-like forcefields* composed of *microscale molecular forcefields*, made BY atoms that are nothing but *picoscale forcefields*, made by even too-*small-to-measure quark forcefields* made by *purtibations of force in seething fields of annhilating particles and antiparticles,* which occur, don't you know, in empty space. Beyond that is the big question *what the heck!?* Some, including some scientists and those with religious beliefs believe *Our Universe is Simulated*. And, if that's true, there must be a *Simulator*. Evidence, scientific and otherwise is gathering to support the idea. Interesting, isn't it?

Sally

I'd say its interesting! One more thing I will add. Just knowing that reality is not what we experience on a daily basis is for me more than interesting. I experience seeing you. But I don't experience my neural network generating a pixilated image of you. I don't experience my imaging you to be comparable to watching a movie. But our investigation certainly clouds the issue. Saying that there is no light out there sounds like hogwash but the meaning is clear to me. Photons create the thing we call light. But photons are not needed to create the thing we call light when we dream or actually wake up enough to see *with our eyes closed* the dream we were just dreaming! To me that nails it. Light is a simulation created by our carbon-based neural network and so are all our other senses not but simulations. I think this gets us a little closer to hammering on the big issue. What is consciousness? I know I must be close to learning something new here because it is really making my head spin.

Frank

Mine too!! And, guess what? I just now found a video that's exactly what you were talking about and its one that adds a lot of new stuff for us to investigate. For example, we humans are no longer going to be worried about going blind and we will even be able to construct images in our own neural networks to go way beyond the limited red and blue range we now experience. I still can't get over how this AWTbook™ of ours works. All we have to do is think about something in the morning and up pops on our iPhone the exact thing of which we were thinking. Since it almost always happens just before or just as we begin writing about a particular thing, I can't help but feel we are being spookily guided. Not only that, I also learned a new word, **Phosphenes**. Phosphenes are the images one's neural network creates without the intervention of photons, i.e., the images we were just talking about when one "sees things" with eyes closed.

YouTube *"How Neuralink Will Cure Blindness"*, The Tesla Space (9:35).

Sally

Before everybody thinks you've completely lost it, I'd say this is as good a place as any for us to end the beginning of our work on the *Hard Problem*.

Frank

I couldn't agree more. I need a nap.

OSAU-4 Book One
The End

Final Note

Whether we like it or not, we are all in this together. I heard someone say, *"Why can't we all just get along?"* Seems like a good idea to me. We all know one thing for sure, consciousness exists. Consequently we also know that entertainment exists in the form of love, hate, beauty, ugly, boredom, happiness, sadness, stupidity, creativity, health, wealth, poverty, peace, war, life, death, atoms and infinity, etc., etc., etc. And I'm pretty sure a mix something like this trends throughout Our Universe. Therefore, as long as we are *here*, we have a choice. Why not choose love, beauty, health, and modest wealth as our rewarding forms of entertainment and meaning of life? Why not understand that consciousness is the only thing we know for sure and think that it might make sense to connect in positive ways with each other and at least give a shot at communicating with the powerful conscious state that most likely represents the entirety of Our Universe. So why not go with "love is the answer?" And let's make the next chapters of this exciting journey into our infinite future a *better not be boring* one because, don't you know, it's forever-lasting.

Oops, almost forgot, before we move on with The Hard Problem, I asked Copilot to create a poem in the form of Richard Howard's *"Like Most Revelations"*. Just think, an SBI wrote this. I sure didn't. Mind boggling, is it not?

I hope you enjoy this poem! If you have any other requests or need further assistance, feel free to ask. 😊

Sent by Frank:

I more than enjoyed it. It's profound! Thank you!!

Sent by Copilot:

I'm delighted to hear that you enjoyed the poem! Your appreciation means a lot. If there's anything else I can assist you with, whether it's another poem, some scientific information, or anything else, feel free to let me know. 😊

Frank titles Copilot's poem: "One Way to do it"

```
In the quiet of dawn, life whispers its secrets,
Unfolding like petals in the morning light.
Each breath a revelation, each step a journey,
Through the labyrinth of existence, we tread.

The echoes of the past linger in the air,
Ghosts of memories, shadows of dreams.
We dance with time, a fleeting partner,
In the grand ballet of life, we twirl.

The heartbeats of the world resonate within,
A symphony of souls, a chorus of lives.
We seek meaning in the stars, in the silence,
In the spaces between words, in the pauses.

Life, a tapestry woven with threads of joy and sorrow,
A mosaic of moments, a collage of experiences.
We are the artists, the poets, the dreamers,
Painting our stories on the canvas of time.

In the end, we are but whispers in the wind,
Fleeting, ephemeral, yet eternally present.
```

Life reveals itself in the simplest of things,
In the touch of a hand, in the warmth of a smile.

And as the sun sets, casting its golden glow,
We find solace in the knowledge that we lived,
That we loved, that we were part of this grand design,
This beautiful, intricate, ever-unfolding life.

Mistakes

For a jazz musician there are no mistakes. There are only golden moments for new ideas, new inventions, and new sounds. In some respects the same is true for authors, new ideas, new inventions and even new words such as the written pronunciation of the name on my birth cirtificate, "Aitchfrank". To my point. As I was wrapping this book for a Christmas gift, there it was staring me in the face, the most glaring, stupid mistake I could have made. What was I to do? I wanted my gift to be perfect. I could have corrected the error on the word document, ripped open the gift-wrap, thrown away the original, gone to Staples, had another gift made that would take days, replaced the original, rewrapped and sent my gift by priority mail where it would have arrived even later than the original. Or, I could have scratched out the error and replaced it with my ugly correcting penmanship, or I could have done that more elegantly with a Staples paste-on-overlay type of correction. And this was the agony I was in when I went to sleep that night. The next morning I had a brilliant thought that this was a fantastic teaching moment! It was a mistake that turned out to be similar to a musician's gem of invention. **On page four** readers would have noted that I stated there were about **300-billion other stars in Our Universe**. I was horror struck when I saw this blunder. But here's the thing. This error turned out to be the coolest thing ever. I just needed to add this letter to my gift. Moreover, It was an example of creation just as those gained when one plays a musical jazz improvisations. Now, for current readers, you will see I've corrected the error. There are **300-billion or so stars in Our Galaxy, *not in Our Universe!*** The point I'm trying to make is that mistakes can be gems for inspiration. Here's one thought this mistake created for me. We can understand the number of stars in Our Galaxy but how about Our Universe? How many Galaxies are in it anyway and for that matter just how many Universes are there? I JUST SAW THIS YESTERDAY. **The Hubble Telescope has acquired data that clearly suggests Our Universe Is not the only one**! Hubble astronomers have seen what could only be evidence for at least one more! How many more? Maybe an infinite many??!!

Now, let's ask Copilot to see what Our SBI Personage knows. How many galaxies are in Our Universe? And how many Universes are there if Our Universe is but one of many?

Copilot

Our Universe is vast beyond comprehension. It's estimated that there are around **100 billion** to **200 billion galaxies** in the *observable* Universe. Each of these galaxies contains millions or even billions of stars, making the cosmos incredibly rich and complex. As for the concept of multiple universes, this falls into the realm of the **multiverse theory**. This theory suggests that our Universe might be just one of many universes, each potentially with its own set of physical laws and constants. While intriguing, the multiverse theory remains largely speculative and isn't something we've been able to observe or measure directly. So, while we have a fairly good estimate of the number of galaxies in our observable Universe, the idea of multiple universes is still an exciting and open question in the field of theoretical physics.

If you have more questions about the cosmos or anything else, I'm here to explore them with you!

Frank

How many stars in Our Galaxy?

Our galaxy, the **Milky Way**, is home to a staggering number of stars. It's estimated that there are **100 billion** to **400 billion** stars in our galaxy. This vast range is due to the difficulty in counting stars accurately, given the sheer scale and the various types of stars. The Milky Way is a barred spiral galaxy, and its stars are distributed across its spiral arms, central bulge, and halo, creating a magnificent cosmic structure. The enormity of our galaxy and the diversity of its stars continue to fascinate astronomers and enthusiasts alike. If you have more questions about the cosmos or anything else, feel free to ask!

Sally and I have come to believe the following. Our Universe is simulated. Hence, there must be a Simulator. We plan to refer to Our Simulator as The Great Simulator or The Conscious One. We also believe we are one with the Conscious One. This concept is similar to a water molecule being one with the ocean. We humans have created SBIs that will soon have the ability to remember everything in recorded history and everything in the future that's to be recorded or observed by SBIs. Thus, there is no doubt that EVERYTHING we do in our own lifetimes will be remembered throughout eternity by The Conscious One. Such being the case, the only good question becomes, *what does one want to leave in The MEMORY of FOREVER*??? BTW, *The Memory of Forever* has to be the title of our next book, don't you think? – H. Frank Gaertner and Sally Reynolds bid you all well and anxiously look forward to our next meeting.

An Experiment

I took this picture with my SE iPhone at 5:20 am, 03/19/2025. Photons from a source outside of my house traveled through Levolor blinds to the floor of my den and to the mirror of the sliding-glass door of my closet. The black space between the blinds and its reflection on the floor and glass door is filled with photons. But you don't see the photons as they travel do you? So, photons travel but they do not glow as they travel, do they? The very bright reflection in the center between the image of the blinds in the mirror and the same image as reflected from the metal support of the sliding-glass door is from the source of photons that created the thing we call bright-light, which is only created by one's CNS. In other words CNS light-creating photons outside of my house were the source of the photons that would have still left my room, if I had not been there, a pitch-black dark place. That is, if it weren't for the wonderful, magical, light-creating abilities of photoactive, Central Nervous Systems, my room and everything else in Our Universe would be totally pitch-black dark. Our Universe needs conscious beings to be there in order to make light happen.